Common Native Plants of Central Florida

Robert H. Stamps
and
Loretta N. Satterthwaite

Editors

Copyright 1994
by
Tarflower Chapter, Florida Native Plant Society

Second printing, 1995

ISBN 1-885258-03-8

Printed on recycled paper

ACKNOWLEDGEMENTS

Thanks are extended to all who helped make this publication possible.

Writers
Bob Stamps
Dick Deuerling
Bill Grey
Kathy Hale
Dan Homblette
Sam Hopkins
Jim Lee
Mike Martin
Loretta Satterthwaite

Illustrators
Bill Bissett
Andrea Burnap
Ann Chase
Carolyn Cohen
Tom Daly
Paul Davis
Vicki Ferguson
Steve Harrison
Patti Hart-Eldridge
Mary Kutz
Peggy Lantz
Jim Lee
Donna Nelson
Loretta Satterthwaite
Bob Stamps
Donna Stringfellow
Rani Vajravelu

Other contributors
Mike Mingea, Executive Director, Florida Native Plant Society
Environmental Horticulture Department, University of Florida, Gainesville (picture glossary)

Layout and Typesetting Courtesy of
Irmo Images, 6330 Plymouth-Sorrento Road, Apopka, FL 32712

Table of Contents

Preface

Common Native Plants of Central Florida — Trees and Shrubs is intended as a guide to use in the field as well as for aiding in determination of what native plants might be suitable for a particular landscape use. The plant pages include the more often used common names, the genus and species names (by which the plants are listed in alphabetical order), and the family to which each plant belongs. On each plant page, some characteristics are listed which should help distinguish that species from most other species; many of the pages then use the numbers in the "identifying characteristics" to refer to drawings which attempt to show that characteristic. Please remember that the drawings may be from the wild or from a cultivated plant. Sizes are generally given as ranges, often with a maximum, but a particular specimen may surpass the maximum when growing conditions are optimum; therefore, no absolutes are intended by the authors.

The book has been a *completely volunteer project* which has taken several years to complete. Several of the volunteers were called on time and time again to do "just one more thing" in order to bring this book to fruition and we thank them as well as each and every volunteer who has contributed to this effort. We would also like to thank the Tarflower Chapter of the Florida Native Plant Society (members and board) for their patience and support throughout the project.

We hope that you will find this book to be of value to you and that you make good use of it by starting — if you have not already done so — to appreciate native plants in the wild and to use them appropriately in your landscaping. We hope that an appreciation of native plants will lead individuals to work to preserve native habitats in Florida and around the world.

Bob and Loretta

Some Central Florida Habitats

Plants and animal species are adapted to live in specific environments called **habitats or communities**. Some of the physical factors that determine environments are fire, soil, water, radiation, temperature and topography. An environment is also defined by biological factors — the plants and animals that are present. Central Florida is blessed with a diversity of natural plant communities. These communities can be categorized in many ways. We will divide communities broadly into three groups — upland, wetland and aquatic habitats.

Uplands are not regularly flooded and their soils are infrequently saturated with water. Wetlands are areas frequently inundated with freshwater. Aquatic habitats such as lakes and streams are covered with water much of the time. Listed below are descriptions of some of the common upland, wetland and aquatic habitats common in central Florida.

UPLANDS

Scrub — Old dunes commonly with deep, excessively drained soils. Because the soils do not retain moisture well, this habitat is called xeric (little moisture available to support plant life). Despite the unglamorous name, scrub areas are important because they are home to some of Florida's rarest plants and often serve as one of the prime water recharge areas for underground aquifers (about 91% of the drinking water in Florida comes from our aquifers). Plants inhabiting these dry, desertlike (xeric) environments often have adaptations such as leathery leaves with waxy coatings which reduce water loss. In spite of the dry nature of these areas, fires are infrequent. Common plants of scrub areas are sand pine, rusty lyonia, myrtle oak, rosemary and saw palmetto.

Sandhill — Rolling hills with deep, sandy, well-drained soils. Fires are frequent (occur annually or every few years) in this xeric habitat so these areas are often dominated by longleaf pine and/or turkey oak, species adapted to frequent fires. Young longleaf pines go through a "grass stage" in which they develop deep roots and thick stems but remain short (look like tufts of grass) and thereby avoid the more intense heat of fires. After several years in the "grass stage", longleaf pine stems elongate rapidly, propelling the growing point of the plant to a height that reduces the chances for fire damage. Turkey oaks spread by underground runners that are insulated from fires by the soil. Wiregrass is a common understory plant in sandhills.

Xeric Hammock — Characterized by deep sandy soils with fires absent or rare. Live oaks — with limbs covered with resurrection fern, orchids and bromeliads — are common overstory plants in this habitat. These hammocks are ideal sites for development and therefore are disappearing rapidly.

Upland Hardwood Forest — This habitat is more mesic (moisture moderately available) than the preceding ones due to the presence of clay in the soil which reduces water drainage past the root zone of the plants that occur there. Fire is rare in this habitat. Magnolias, oaks, hickories, beeches and hollies are typical of these areas.

Flatwoods — This common habitat is characterized vegetatively by slash and longleaf pine. Fires are frequent to occasional. Common understory plants include saw palmetto, gallberry and wiregrass.

WETLANDS

Hydric Hammock — Lowland, mesic-hydric habitat. Characteristic vegetation includes water oak, cabbage palm, bays, red maple and gums. Fires are rare in this habitat.

Freshwater Swamp — Low, wet areas characterized by cypress, gum and bay trees. Freshwater swamps are flooded at least part of the year. Fires are rare or absent and soils contain more organic matter than the upland habitats described above. Willows and ferns are common in this habitat.

Wet Prairie — These flatlands are seasonally inundated with water and are characterized by maidencane, wiregrass, rushes and other low-growing plants. Trees are few due to frequent fires and periodic inundation.

AQUATIC

Lakes and Ponds — Still (as opposed to moving) waters occurring in depressions. Water supply is usually from surface inflows, high water tables, or direct connections to the aquifer (sinkholes) in central Florida. Subsurface water loss is often restricted by clay, limestone and/or organic matter. Vegetation consists of emergent plants occurring around the edges of these bodies of water and floating/submerged aquatic vegetation living in the water. The emergent vegetation is composed of plants adapted to wet soils such as cypresses and tupelos.

Streams — Flowing waters bounded by banks. These habitats may have water flowing year-round (spring-fed) or intermittently (depending on rainfall). Needle palms, red maples and cypress are typically found along the banks.

Picture Glossary

Leaf Terminology I Blade Margin

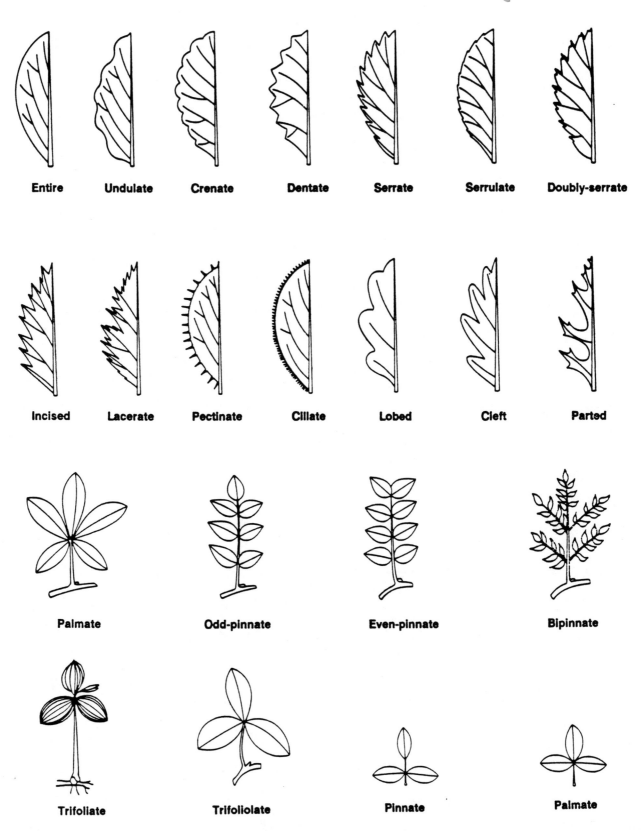

Entire   **Undulate**   **Crenate**   **Dentate**   **Serrate**   **Serrulate**   **Doubly-serrate**

Incised   **Lacerate**   **Pectinate**   **Ciliate**   **Lobed**   **Cleft**   **Parted**

Palmate   **Odd-pinnate**   **Even-pinnate**   **Bipinnate**

Trifoliate   **Trifoliolate**   **Pinnate**   **Palmate**

Picture Glossary

Leaf Terminology II Blade Shape

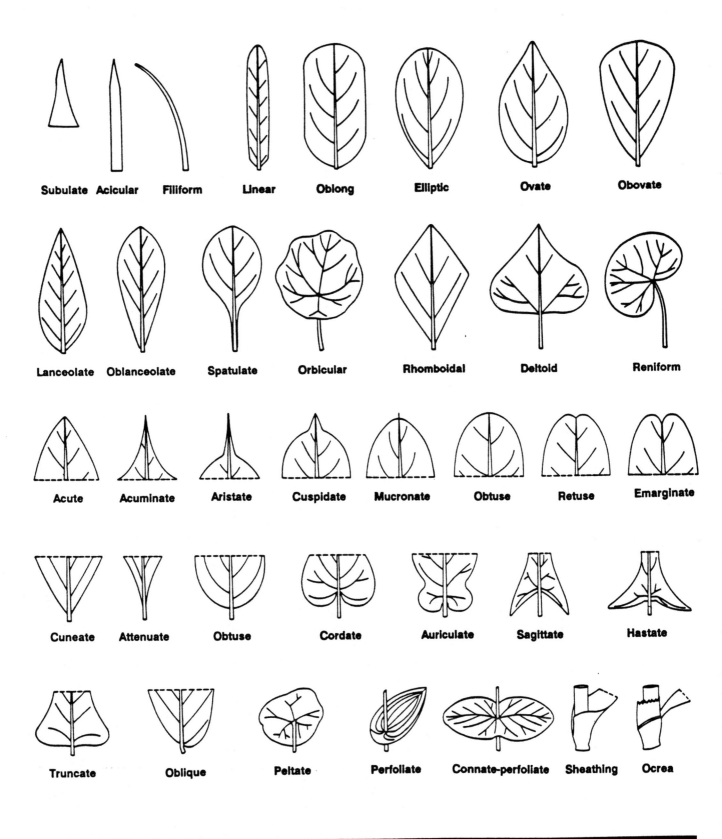

Subulate Acicular Filiform Linear Oblong Elliptic Ovate Obovate

Lanceolate Oblanceolate Spatulate Orbicular Rhomboidal Deltoid Reniform

Acute Acuminate Aristate Cuspidate Mucronate Obtuse Retuse Emarginate

Cuneate Attenuate Obtuse Cordate Auriculate Sagittate Hastate

Truncate Oblique Peltate Perfoliate Connate-perfoliate Sheathing Ocrea

Picture Glossary

Inflorescenses

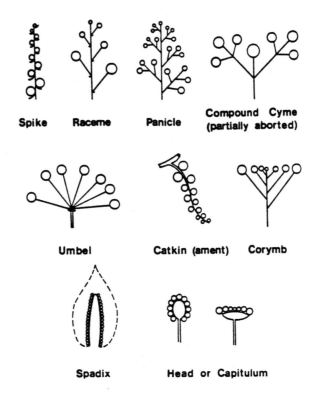

Spike **Raceme** **Panicle** **Compound Cyme (partially aborted)**

Umbel **Catkin (ament)** **Corymb**

Spadix **Head or Capitulum**

Palm Terms

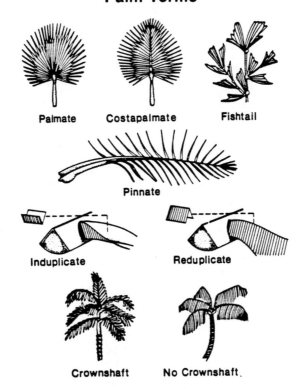

Palmate **Costapalmate** **Fishtail**

Pinnate

Induplicate **Reduplicate**

Crownshaft **No Crownshaft**

Trees and Shrubs

(SOUTHERN) RED MAPLE
(*Acer rubrum*)
Family: ACERACEAE

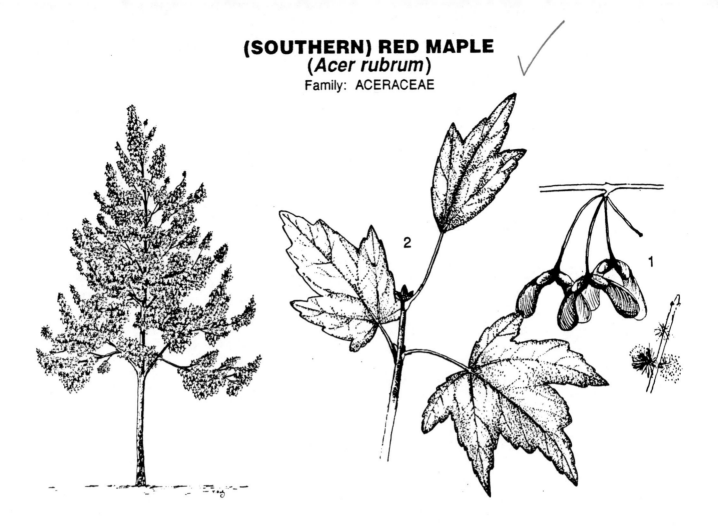

IDENTIFYING CHARACTERISTICS (see below for details): 1) red seed-bearing samaras in the spring; 2) leaf shape and color — palmate, 3- to 5-lobed, with red stalks.

DESCRIPTION	Small to large deciduous tree with well-developed trunk and spreading crown.
GROWTH RATE	Fast, especially in moist soils.
SIZE	Eighty to 90' tall; trunk 2' in diameter.
LEAVES	Opposite, simple, 2 to 5½" long, often as broad as long; 3 to 5 deep sharp lobes, margins toothed, stalks red; thin, deep green above and paler beneath.
FLOWERS	Red to yellow, bloom very early in winter and spring before leaves appear.
FRUIT	Shiny green seed maturing in spring, borne in 1" or more long, paired red papery wings (samaras).
BARK	Brownish-gray with shallow furrows; branchlets usually reddish.
RANGE	Eastern U. S., in every county of Florida.
HABITAT	Common on moist to wet ground, but will grow on much drier sites. Shade or full sun.
LANDSCAPE USAGE	The red maple is an excellent landscape plant with a fast rate of growth. Used as a specimen, street and woodland tree. Provides summer shade and fall color and because it drops its leaves in the winter, it can be used in energy-conserving landscape designs.
HUMAN EDIBILITY	The seed may be boiled like a pea, which it resembles. The sap, while not so good as that of the sugar maple (*Acer saccharum*), may be used for syrup and sugar.
REMARKS	Sensitive to wounding, but a beautiful tree which should be used more widely. Provides food and nest-building materials for wildlife.

GROUNDSEL TREE, SALTBRUSH, SEA MYRTLE
(*Baccharis halimifolia*)
Family: ASTERACEAE or COMPOSITAE

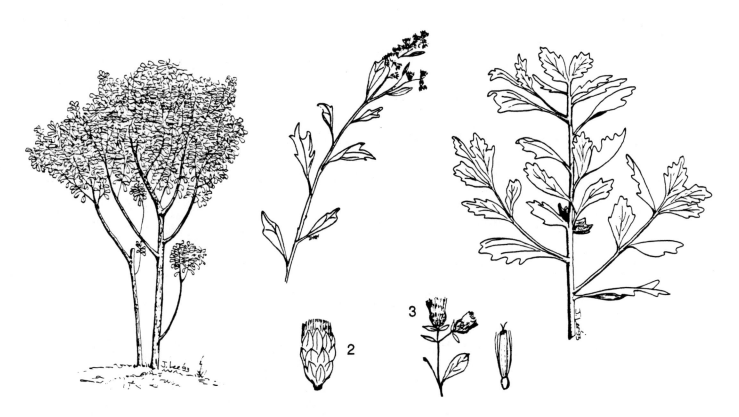

IDENTIFYING CHARACTERISTICS (see below for details): 1) grayish-green leaves — with conspicuous amber-colored glands on underside; 2) flowers — silvery paint brushes; 3) fruitheads — conspicuous, fuzzy, white.

DESCRIPTION	Shrub or small, much-branched, spreading tree.
GROWTH RATE	Moderately fast.
SIZE	Twelve to 16' tall and 4–5" diameter trunk with narrow conical crown.
LEAVES	Alternate, simple, persistent, $\frac{1}{2}$ to 3" long and $\frac{2}{5}$ to $1\frac{1}{4}$" wide; elliptical to obovate, usually a few coarse teeth toward the short-pointed tip and, easily seen, amber glandular dots on underside; dark gray-green and dull above and paler beneath.
FLOWERS	Dioecious. Tiny, in greenish bell-shaped heads, less than $\frac{1}{4}$" long, resembling silvery paint brushes; from August to November.
FRUIT	Brushlike, whitish head $\frac{1}{2}$" long and wide; seeds with tuft of whitish hairs to $\frac{2}{5}$" long.
BARK	Brownish-gray, shallowly furrowed.
RANGE	Occurs along the Atlantic and Gulf coasts and throughout Florida.
HABITAT	Moist soils, roadside ditches, open woods, marshes, and disturbed areas. Tolerant of drier soils.
LANDSCAPE USAGE	Especially suitable for naturalistic landscapes, tolerant of saltwater spray. Masses of fuzzy, white fruitheads on female plants make showy display, but seeds spread by the wind may germinate where they are unwanted and require weeding.
HUMAN EDIBILITY	None
REMARKS	One of only two members of the composite family native to North America that may attain tree-size.

AMERICAN BEAUTYBERRY, BEAUTYBUSH
(*Callicarpa americana*)
Family: VERBENACEAE

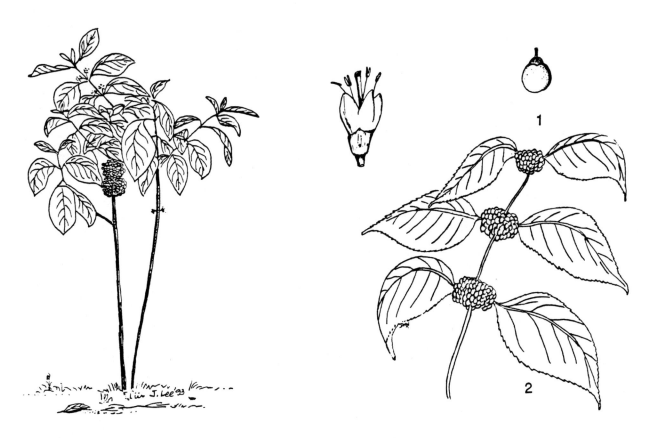

IDENTIFYING CHARACTERISTICS (see below for details): 1) fruit — fleshy, violet or purple, in clusters encircling the stems; 2) leaves — opposite, long, ovate/elliptical, finely-toothed.

DESCRIPTION	Deciduous shrub with gray hairy twigs and an open growth habit.
GROWTH RATE	Moderate.
SIZE	Three to 8' high with slender spreading branches.
LEAVES	Opposite, simple, 3 to 6" long and 1 to 3" wide; ovate to elliptical, pointed ends, finely toothed; slightly rough beneath.
FLOWERS	Small, tubular, clustered in axils of the leaves, pinkish to bluish, mainly late spring and summer.
FRUIT	Fleshy, bright violet or purple, round $\frac{1}{8}$–$\frac{3}{16}$" in diameter, thin-skinned; occur in showy clusters, $\frac{3}{4}$" or more in diameter, in late summer and persist into winter.
BARK	Brown with numerous lenticels.
RANGE	Southeastern U.S., throughout Florida.
HABITAT	Pinelands and hammocks. Grows well in most soils.
LANDSCAPE USAGE	Specimen plants. Mass planting under light shade such as group of pine trees, but tolerant of full sun. Easily propagated and transplanted.
HUMAN EDIBILITY	Fruit edible raw but lacking flavor and astringent; however, one fruit will get the saliva running if your canteen is empty. Jelly can be made from fruit.
REMARKS	Fruit is eaten by many bird species. Fruit-covered stems of beautyberry are stripped of leaves and used in floral arrangements.

IRONWOOD, AMERICAN HORNBEAM, BLUE BEECH, WATER BEECH
(*Carpinus caroliniana*)
Family: BETULACEAE

IDENTIFYING CHARACTERISTICS (see below for details): 1) bark of older trees — smooth, flexed-muscle look; 2) fruit — ovoid, coarsely-ribbed nutlet with 3-lobed leaflike bracts.

DESCRIPTION	Small to medium-sized deciduous, monoecious hardwood with smooth flexuous gray bark.
GROWTH RATE	Slow.
SIZE	Mostly 20 to 30' tall; average trunk diameter 4 to 8".
LEAVES	Alternate, simple, elliptic to ovate, doubly serrate margins with rounded base and acuminate tip; $3/4$–4" long, $3/4$–2" wide.
FLOWERS	Inconspicuous, occur as unisexual catkins on the same tree, male flowers in long, 4", drooping catkins and female flowers, with bracts close beneath them, are in short, terminate catkins.
FRUIT	Ovoid, coarsely-ribbed nutlet, $1/5$–$1/4$" long, occuring in "leafy" clusters formed by 3-lobed bracts at base of each nutlet. Middle lobe of bract can be 1" wide and $1\frac{1}{2}$" long.
BARK	Blue-gray and smooth; trunk and major branches of older specimens have a distinct appearance often resembling flexed muscles.
RANGE	Generally throughout eastern U.S. into central Florida, west to east Texas.
HABITAT	Stream banks and floodplain communities associated with rich soils, or in mesic hardwood forests; a characteristic subcanopy species.
LANDSCAPE USAGE	Specimen tree.
HUMAN EDIBILITY	Small nutlet may be parched (placed in skillet, shaken over heat enough to warm) and eaten, but too small to be worthwhile except for survival.
REMARKS	Wildlife food source. Wood is strong, used for tool handles.

PIGNUT HICKORY
(*Carya glabra*)
Family: JUGLANDACEAE

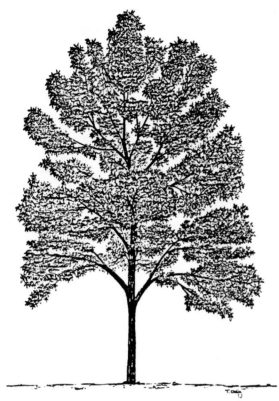

IDENTIFYING CHARACTERISTICS (see below for details): 1) bark — diamond pattern; 2) leaves — pinnately compound, usually with 5 leaflets; 3) fruit — thick-shelled nut, not ribbed.

DESCRIPTION	Medium to large deciduous tree of hardwood forests with pear-shaped fruit.
GROWTH RATE	Slow.
SIZE	Attains heights of 50–75' or greater; trunk up to 3' in diameter.
LEAVES	Alternate, pinnately compound, 6 to 12" long with 5–7 leaflets that are variable in shape — lanceolate, elliptic, oblong or ovate — with asymmetrical bases; margins finely serrate, mature leaves glabrous above, varyingly pubescent to glabrous below.
FLOWERS	Male flowers are drooping catkins, female flowers are 2–10 flowered racemes.
FRUIT	Thick-shelled nut not ribbed, olive-green to dark brown husk, 1 to $2\frac{1}{2}$" long, pear-shaped.
BARK	Light to dark gray; smooth on young trees, shallowly furrowed bark with low interlacing ridges forming a diamond pattern on older trees.
RANGE	Eastern Massachusetts west to eastern Missouri south to southern Alabama and south-central Florida.
HABITAT	Hammocks and rich woods with fertile sandy or clayey soils, dry and moist uplands.
LANDSCAPE USAGE	Specimen or shade tree, park and street plantings. Falling debris (stems, nuts) is likely, especially with larger specimens. Taproot on this tree makes it hard to transplant successfully.
HUMAN EDIBILITY	Nutritious, sweet-meated nut, but sometimes can be bitter.
REMARKS	Wildlife food source. Will tolerate high winds, but is not salt tolerant. Green wood used to flavor meats during cooking. Wood often made into tool handles. Scrub hickory (*Carya floridana*) is found only in central Florida; this small (10–20') tree, which is closely related to pignut hickory, occurs on droughty soils (scrub and dry pinelands) and has smaller ($<\frac{5}{8}$") nuts.

BUTTONBUSH, HONEY-BALLS
(*Cephalanthus occidentalis*)
Family: RUBIACEAE

IDENTIFYING CHARACTERISTICS (see below for details): 1) flowers — white, terminal or axillary pincushions; 2) fruit — rough-surfaced brown balls splitting into nutlets.

DESCRIPTION	Deciduous woody shrub, rarely a small tree; long-lived perennial, older portions of plant often regenerating new branches.
GROWTH RATE	Moderately slow.
SIZE	Generally up to 7' in height, may reach 25' or more; 4" trunk diameter.
LEAVES	Opposite or whorled, ovate to elliptic-lanceolate, pinnately veined; 2–6" long, 1–4" wide, margins entire, apex cuspidate or acute. Leaf surface pubescent to glabrous.
FLOWERS	Showy inflorescence in globose heads 1" or greater in diameter, resembling pincushions. White, terminal and axillary. Flowering period - spring to fall.
FRUIT	Small angular individual fruits combine to form rough-surfaced brown balls up to 1" in diameter; may persist through winter.
BARK	Young twigs are reddish-brown, becoming gray to brown and deeply furrowed with age.
RANGE	Found throughout the southeastern U.S. in appropriate habitats; common throughout Florida.
HABITAT	Common shrub of wetland habitats including river and stream margins, swamps and shallow marshes.
LANDSCAPE USAGE	Handsome ornamental. Good in full sun at water's edge or in moist soil. Rejuvenated by cutting to the ground every few years.
HUMAN EDIBILITY	None. Toxic.
REMARKS	Seeds are utilized by waterbirds and other wildlife as a food source. Foliage is poisonous to livestock.

(SANDHILL) ROSEMARY
(*Ceratiola ericoides*)
Family: EMPETRACEAE

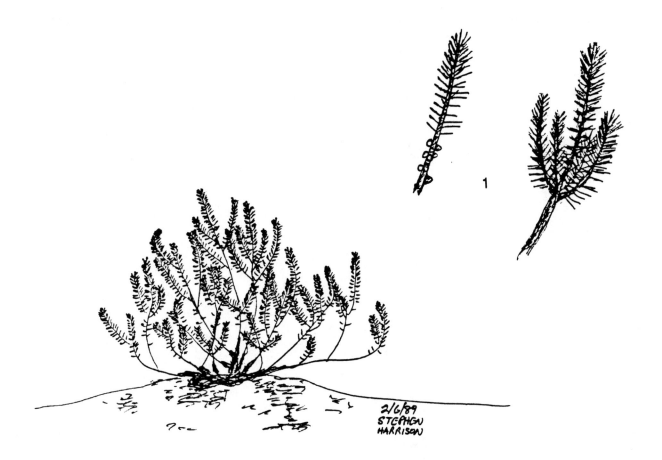

IDENTIFYING CHARACTERISTICS (see below for details): 1) leaves — almost tubular, needlelike; 2) densely branched formation.

DESCRIPTION	Dioecious, evergreen shrub; densely branched.
GROWTH RATE	Slow.
SIZE	Two to 7' tall and 3 to 5' broad.
LEAVES	Sessile, almost tubular, appearing needlelike, alternate or whorled, linear and small ($^1/_5$–$^1/_2$" long); acute apex, aromatic.
FLOWERS	Sessile in leaf axils, reddish-brown or yellowish; appearing spring to fall.
FRUIT	Red or yellow, two-seeded drupes, imperfectly spherical, less than $^1/_5$"; maturing in late fall.
BARK	Rough, brownish.
RANGE	Occurring from South Carolina to Mississippi; common in north and central Florida in appropriate habitat.
HABITAT	Occurrence generally restricted to deep sandy ridges, or xeric oak or sand pine communities.
LANDSCAPE USAGE	Specimen plant in xeric habitats. Do not try to transplant from the wild.
HUMAN EDIBILITY	None.
REMARKS	Sometimes used for fragrance. Only species in its family native to southeast U. S.

FLOWERING DOGWOOD
(*Cornus florida*)
Family: CORNACEAE

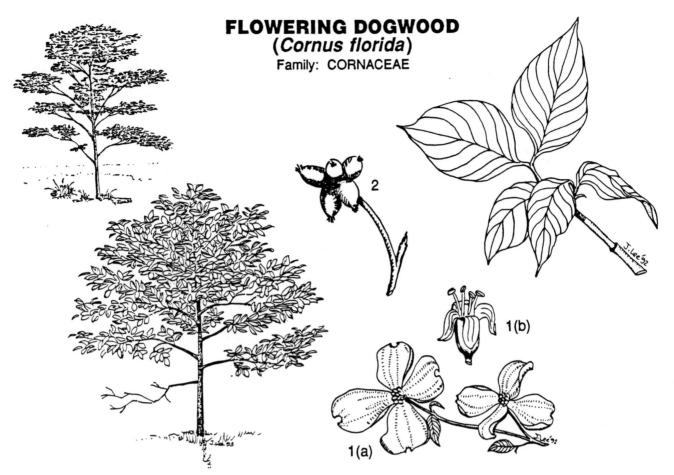

IDENTIFYING CHARACTERISTICS (see below for details): 1) flowers — (a) large, showy, white, notched bracts surrounding (b) true flowers; 2) fruit — red, egg-shaped, in clusters; 3) when leaves of this genus are gently torn apart crosswise, silvery veins remain intact, holding parts together.

DESCRIPTION	Small, leafy, deciduous tree with a rounded, bushy crown that may have a spread wider than the height.
GROWTH RATE	Slow to moderate.
SIZE	Fifteen to 40' tall; trunk 6 to 18" diameter.
LEAVES	Opposite, deciduous, simple, 3–6" long, 1½–2½" wide; ovate to broadly elliptic with pointed apex and 6–7 vein pairs; turn scarlet in the fall. When leaves are gently pulled in half crosswise, veins appear as silvery threads holding leaf blade parts together.
FLOWERS	Large showy white, rarely pink, notched "flowers" are actually bracts surrounding tight clusters of inconspicuous greenish-white true flowers. Appear in early spring, prior to or simultaneous with leafing.
FRUIT	Bright scarlet egg-shaped fruits, ⅓–½" in diameter, with 4-lobed persistent calyx; grouped in clusters.
BARK	Young twigs have a waxy bloom and are green; smooth, gray-brown bark of young trees divides into small, scaly blocks at maturity.
RANGE	Maine to Illinois and Kansas, south to central Florida and Mexico.
HABITAT	Prefers moist, rich soils. Forest edges and clearings; common along fence rows.
LANDSCAPE USAGE	Excellent specimen plant. Full sun to semishade on well-drained soils; cannot tolerate extended drought. Many cultivars available.
HUMAN EDIBILITY	None.
REMARKS	Desirable for spring flowers and fall colors. Fruit eaten by squirrels and birds. Often harvested to excess for its beautiful hard wood for making small tools and wood engraving. Spot anthracnose (*Colletotrichum* sp. and *Elsinoë corni*) can occur in the fall and knock leaves off trees; however, since the leaves are deciduous, there is usually no need to treat with fungicide.

COMMON PERSIMMON, POSSUMWOOD
(*Diospyros virginiana*)
Family: EBENACEAE

1

IDENTIFYING CHARACTERISTICS (see below for details): 1) fruit — smooth, yellow-orange when unripe; puckered, dark orange to brown when ripened; 2) leaves – commonly, blackish blemishes on upper surface; lower surface whitish; 3) twigs — terminal buds often abort, causing twigs to be bushy and to zigzag back and forth.

DESCRIPTION	Medium-sized deciduous tree bearing large, edible fruit and inhabiting a wide range of sites.
GROWTH RATE	Slow to moderate.
SIZE	Generally to 60', may attain heights of 100' in hammock environments; trunk up to 2' diameter.
LEAVES	Alternate, deciduous, simple, $2\frac{1}{2}$–6" long, dark green and shiny above, pale beneath; ovate to elliptic, leaf edge smooth, sharp pointed; broad wedge-shaped base.
FLOWERS	Flowers inconspicuous, $\frac{2}{5}$" in clusters (male), $\frac{3}{5}$" solitary (female); generally on separate plants.
FRUIT	Several-seeded globose or somewhat flattened berry with persistent sepals. Three quarters to $1\frac{1}{2}$" in diameter, yellow to orange-brown. Matures in late summer.
BARK	Dark brown or gray (almost black), deeply checkered with fissures. Twigs often zigzag in appearance.
RANGE	Southern New England west to Kansas, south to east Texas and Florida. Found throughout the state.
HABITAT	Inhabits a wide range of sites with varying moisture regimes including open fields, upland woodlands, pine flatwoods and river bottomlands.
LANDSCAPE USAGE	Fruiting specimen plant. Can be used in parks, naturalized landscapes and yards where fruit drop would not be a problem.
HUMAN EDIBILITY	Fruit edible when mushy ripe — can be baked into a nut bread or pulp can be used for topping on ice cream. Seeds can be roasted and ground for coffee substitute; dried leaves make an excellent tea (similar to sassafras). A good wine can be made from the fruit.
REMARKS	Fruit eaten by an array of animals. Wood used for furniture veneer and small wooden items; excellent carving wood (ebony family). Root is used for grafting for horticultural varieties.

LOBLOLLY BAY
(*Gordonia lasianthus*)
Family: THEACEAE

IDENTIFYING CHARACTERISTICS (see below for details): 1) flowers — 5 showy, white, waxy petals with silky hairs on underside; 2) saw-edge on mature leaves.

DESCRIPTION	Medium-sized, upright, evergreen tree with cylindrical head.
GROWTH RATE	Moderate.
SIZE	Average height 40 to 50', but can reach 90'; average trunk diameter 12 to 15", up to 20".
LEAVES	Persistent, alternate, simple, leathery, smooth, dark green and shining, 4–6" long, elliptic, with somewhat pointed tips, narrowly wedge-shaped bases and margins with very shallow teeth. Turn bright red before falling.
FLOWERS	Showy, white, about 2½" wide, cup-shaped with 5 waxy petals with silky hairs on lower side and many yellow stamens. Flowers May to September.
FRUIT	Capsule, ripening in fall, hairy, woody, ¾" long, ovate, long-pointed, splitting halfway into 5 parts and releasing winged seeds.
BARK	Dark red-brown or gray, thick, roughened by narrow furrows that separate flat-topped ridges. Young twigs are reddish and smooth.
RANGE	North Carolina south to Mississippi and as far south as Lake Okeechobee in Florida.
HABITAT	Commonly found in swamps, ditches and bayheads throughout central Florida. Found in association with sweet bay, sweetgum and red maple.
LANDSCAPE USAGE	Makes a desirable ornamental tree (as a specimen or in groups) when planted under good growing conditions. Requires abundant moisture.
HUMAN EDIBILITY	None.
REMARKS	Lasianthus means hairy-flowered in Latin. Bark has been used for tanning.

DAHOON HOLLY
(*Ilex cassine*)
Family: AQUIFOLIACEAE

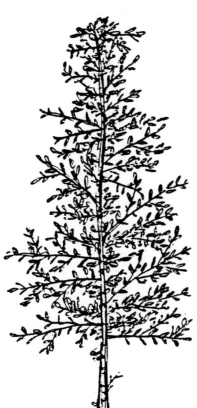

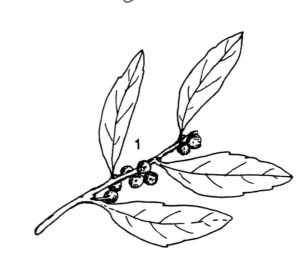

IDENTIFYING CHARACTERISTICS (see below for details): 1) fruit — usually bright red, 1–3 berrylike drupes in axils of leaves in late fall; 2) evergreen tree.

DESCRIPTION	Small evergreen tree with leathery leaves and bright red drupes.
GROWTH RATE	Moderate.
SIZE	Up to about 40', generally 20 to 30' tall and 8 to 15' wide; trunk up to about $1\frac{1}{2}$' in diameter.
LEAVES	Alternate, variable in shape (elliptic, ovate, oblanceolate), 2 to 4" long, medium green, shiny above, pubescent underneath, with a few small teeth near the apex.
FLOWERS	Dioecious, inconspicuous, $\frac{3}{16}$" wide with 4 white petals. Spring flowering.
FRUIT	$\frac{1}{4}$"-diameter, persistent berrylike drupes on female plants range in color from red to yellow. Fruit occur in axillary groups of 1 to 3 and ripen in late fall.
BARK	Grayish, smooth to warty, often covered with lichens. Young stems remain green for 1–2 years.
RANGE	Bahamas, Cuba, and the coastal plain of the southeastern U.S. Throughout all but the southernmost tip of Florida.
HABITAT	Wet habitats. Flatwoods depressions and margins of swamps, ponds and stream banks.
LANDSCAPE USAGE	Specimen or street tree; tall hedge, but tolerates pruning. Fruit on female plants add wintertime color to the landscape. Nice addition to a woodland planting. Requires moist soils. Tolerant of salt spray.
HUMAN EDIBILITY	Non-caffeinated tea can be made from roasted leaves. Fruit is not edible.
REMARKS	Fruit is used as food and plant as nesting and/or cover by wildlife. Propagation by cuttings results in plants of known sex. Hybridizes with other *Ilex* species to produce many notable plants. Grows naturally further south than any other native evergreen, red-fruited holly.

GALLBERRY, INKBERRY
(*Ilex glabra*)
Family: AQUIFOLIACEAE

1

IDENTIFYING CHARACTERISTICS (see below for details): 1) fruit — dull, black berrylike (inkberry) drupe persists through winter; 2) evergreen shrub.

DESCRIPTION	Evergreen shrub common to pine flatwoods and palmetto prairies.
GROWTH RATE	Slow.
SIZE	Average height 3–5', maximum 10'.
LEAVES	Alternate, simple, evergreen, ³/₄–2" long, glabrous, leathery, elliptic to oblanceolate, blunt-tipped; lustrous green above, pale below with scattered glands (usually red); few blunt teeth near tip.
FLOWERS	Small whitish-green; separate pistillate and staminate flowers in axils in winter, spring and summer.
FRUIT	Berrylike, dull, black globose persistent drupe.
BARK	New twigs green, finely powdery pubescent; older twigs glabrous, gray to grayish-brown with lenticels.
RANGE	Coastal plain of Nova Scotia to south Florida and west to northeast Texas. Throughout Florida.
HABITAT	Mesic pine flatwoods and palmetto prairies.
LANDSCAPE USAGE	Foundation, hedge, mass and accent plant for the landscape. Good holly for wet or moist areas. Various cultivars available.
HUMAN EDIBILITY	Very good non-caffeinated tea made from roasted leaves. Fruit is not edible.
REMARKS	Wildlife uses fruit for food and plant for nesting and/or cover. Spreads by creeping stems. Known as a "bee plant".

SOUTHERN RED CEDAR
(*Juniperus silicicola*)
Family: CUPRESSACEAE

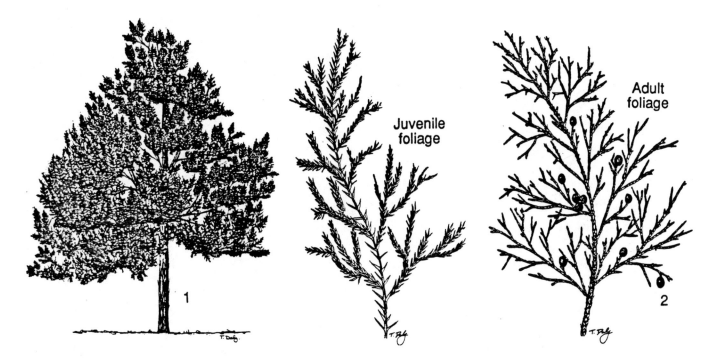

Juvenile foliage

Adult foliage

IDENTIFYING CHARACTERISTICS (see below for details): 1) bark — peels in long stringy shreds; 2) female cones — blue-green, berrylike persist through winter; 3) conspicuous orange, globular "cedar apples" in early spring.

DESCRIPTION	Small to medium dioecious evergreen conifer with a broad, conical form. The spread often equals the height of the tree.
GROWTH RATE	Fast.
SIZE	To 50' in height; trunk up to 2' in diameter.
LEAVES	Evergreen, aromatic, small ($1/_{32}$–$3/_{16}$" long), sharp-pointed, and triangular on adult branchlets; opposite and two-ranked or whorled, often overlapped and so close to twig that it looks quadrangular in cross-section. Juvenile leaves are longer and awl-shaped.
FLOWERS	Not applicable. The gymnosperms bear naked seeds, usually on the scale of a cone (see next item).
FRUIT	Blue-green berrylike female cones which persist through winter. Male plant produces numerous short, cylindrical yellow cones during winter.
BARK	Reddish-brown, very thin, peeling into long, stringy shreds.
RANGE	Coastal North Carolina south to Florida and west to Texas. Widely distributed in Florida — as far south as Sarasota County and into north Florida along the Gulf of Mexico.
HABITAT	Commonly found in areas with moist soils or soils underlain by limestone. Tolerant of a wide range of moisture conditions.
LANDSCAPE USAGE	Accent tree good for soil stabilization, windbreaks and noise abatement plantings; living Christmas trees.
HUMAN EDIBILITY	None.
REMARKS	Fruits used by numerous wildlife species as food and tree is used for nesting and/or cover. Dense stands were once found at Cedar Key and other areas of Florida, especially mosquito impoundment dikes and disturbed shell mounds. In early humid spring weather, orange, globular, gelatinous, spore-producing bodies of a fungal parasite (called "cedar apples") may be conspicuous on tree branches. Bark makes good tinder for making campfires. Demand for the wood for use in pencil making, lumber and aromatic "cedar chests" led to overharvesting. Also used for cut Christmas trees.

SWEETGUM
(*Liquidambar styraciflua*)
Family: HAMAMELIDACEAE

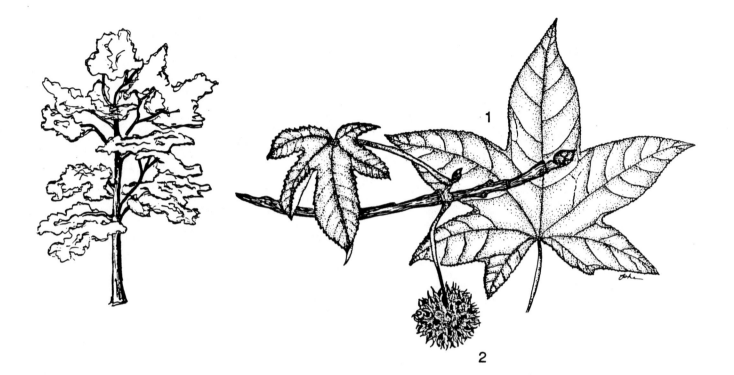

IDENTIFYING CHARACTERISTICS (see below for details): 1) Leaves — alternate, star-shaped with 5 to 7 lobes; 2) fruit — spiny globe 1 to $1\frac{1}{2}$" in diameter.

DESCRIPTION	Large deciduous tree with straight trunk.
GROWTH RATE	Rapid.
SIZE	Eighty to 120' high with 40 to 60' spread; trunk diameter $1\frac{1}{2}$–3', up to 5'.
LEAVES	Simple, alternate, aromatic, 3–7" long, star-shaped with 5 to 7 lobes with saw-toothed edges. Deep glossy green above, changing to red, yellow, or purple in the fall.
FLOWERS	Inconspicuous. Female — pale green, numerous, in tight round heads borne at end of long, drooping stalks; male — greenish-yellow, tightly clustered on stiff spike.
FRUIT	Spiny globe 1 to $1\frac{1}{2}$" in diameter; green when young and brown when mature; opening with many slits to release seeds.
BARK	Light gray, often ridged or warty when young; dark brown, fissured and rough on old trunks.
RANGE	Connecticut to Florida, Missouri, Illinois and south to Mexico.
HABITAT	Moist to wet deciduous woods. Stream banks and low swampy bottomlands.
LANDSCAPE USAGE	The sweet gum can be planted for quick shade; but the fruits can be a nuisance and branches may break off during strong winds. Can be used for highway plantings; however, not highly recommended as a street tree and should only be planted in suitable locations with plenty of room and an abundance of moisture. Cultivars whose leaves consistently change color in the fall are available.
HUMAN EDIBILITY	Sap can be used as chewing gum, as the name implies.
REMARKS	Fruit eaten by birds. Fruit sometimes "decorated" and used as holiday ornaments.

PRIMROSE WILLOW
(*Ludwigia peruviana*)
Family: ONAGRACEAE

IDENTIFYING CHARACTERISTICS (see below for details): 1) flowers — numerous, showy, yellow; 2) many-branched, rapidly growing shrub in wet areas; 3) flowers shatter immediately on picked branch.

DESCRIPTION	Weedy herbaceous or semi-woody, many-branching perennial which grows along wetland edges and in wet drainage ditches.
GROWTH RATE	Rapid growth, aggressive invader with reproduction by seed and rhizome.
SIZE	Three to 5' in height, up to 9'.
LEAVES	Alternate, narrowly lanceolate to lance-ovate (2–6" long), with dense pubescence on both sides.
FLOWERS	Yellow, showy flowers, 4–5 petals $\frac{1}{2}$–2" long in upper leaf axils. Flowers all year.
FRUIT	Numerous brown seeds in a four-angled cylindrical capsule (1").
BARK	Flaky.
RANGE	South Georgia through Florida and the Gulf states to tropical America.
HABITAT	Wetland edges, pond margins and wet ditches.
LANDSCAPE USAGE	Invasive, not recommended since difficult to control.
HUMAN EDIBILITY	None.
REMARKS	Waterfowl utilize the seeds.

RUSTY LYONIA
(*Lyonia ferruginea*)
Family: ERICACEAE

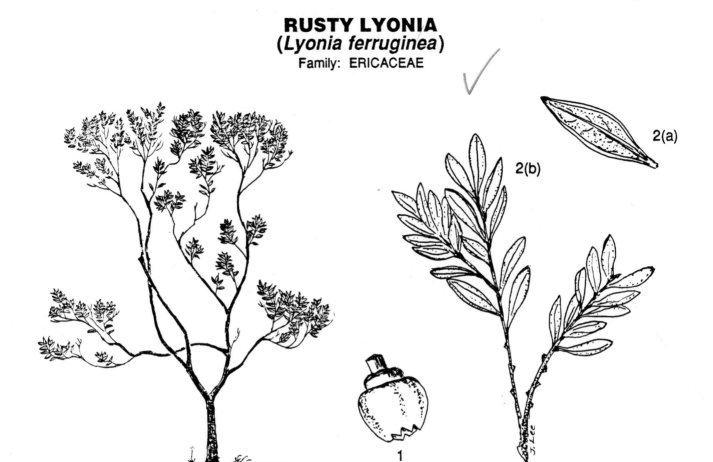

2(a)

2(b)

1

IDENTIFYING CHARACTERISTICS (see below for details): 1) flowers — borne on shoots of previous season; 2) leaves — (a) lower surface covered with reddish-brown scales AND (b) NOT smaller toward end of flowering shoot; 3) bark — reddish-brown with rusty, scalelike particles on crooked branches.

DESCRIPTION	Medium to large evergreen shrub or small tree with broadly arching (crooked) branches.
GROWTH RATE	Slow.
SIZE	Generally 6 to 9' tall, up to 20'; up to 6" in diameter.
LEAVES	Alternate, 1–3½" elliptic to obovate, smooth, downturned (revolute) margins; lower surface covered with reddish-brown scales. Leaves not reduced toward end of flowering shoot.
FLOWERS	White, ⅛" wide urn-shaped in close clusters or short racemes in axils of leaves. Corolla many times longer than calyx. Appear in the spring on previous year's growth.
FRUIT	Ovoid capsule up to ¼" long, with thickened ribs between valves; appearing in early summer, dried brown capsules persist on plant.
BARK	Reddish-brown, twisted. Twigs with rusty scalelike particles.
RANGE	Sporadic from South Carolina throughout southern coastal states; common from north Florida southward.
HABITAT	Xeric (dry) sandhill scrub and dry oak-pine forests, typically in deep sandy soils.
LANDSCAPE USAGE	Specimen plant or in masses.
HUMAN EDIBILITY	None.
REMARKS	Wildlife food source. Ferruginea means rust-colored and refers to lower leaf surfaces. In mature xeric scrub communities, this species may be an important element of the subcanopy. Makes unique walking sticks.

STAGGERBUSH
(*Lyonia fruticosa*)
Family: ERICACEAE

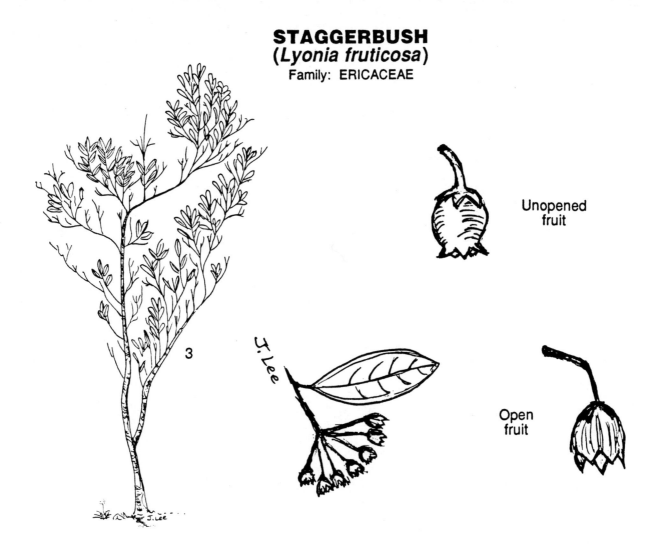

Unopened fruit

Open fruit

J. Lee

3

IDENTIFYING CHARACTERISTICS (see below for details): 1) flowers — borne on new shoots of the season; 2) leaves — conspicuously reduced (smaller) toward end of flowering shoots; 3) shrub with rigidly ascending branches.

DESCRIPTION	Small to medium-sized evergreen shrub with rigidly ascending branchlets.
GROWTH RATE	Moderate.
SIZE	Six feet tall, rarely reaches 10'.
LEAVES	Alternate, to 2" long, stiff, ovate to elliptical, margins not rolled under; leaves conspicuously smaller towards end of flowering shoot. Rust-colored scales/flakiness on under surface.
FLOWERS	White, urn-shaped umbel-like clusters in axils of leaves. Corolla many times longer than calyx. Usually borne on new shoots of the season.
FRUIT	Capsules prominently angled and longer than wide.
BARK	Young stems hairy.
RANGE	South Carolina, Georgia and Florida.
HABITAT	Coastal Plains; pine flatwoods. Sandy, acid soils.
LANDSCAPE USAGE	Use as specimen plant or in mass plantings. Can withstand slightly wetter, more poorly drained soils than *L. ferruginea*.
HUMAN EDIBILITY	None.
REMARKS	Wildlife food source.

FETTERBUSH, SHINY LYONIA
(*Lyonia lucida*)
Family: ERICACEAE

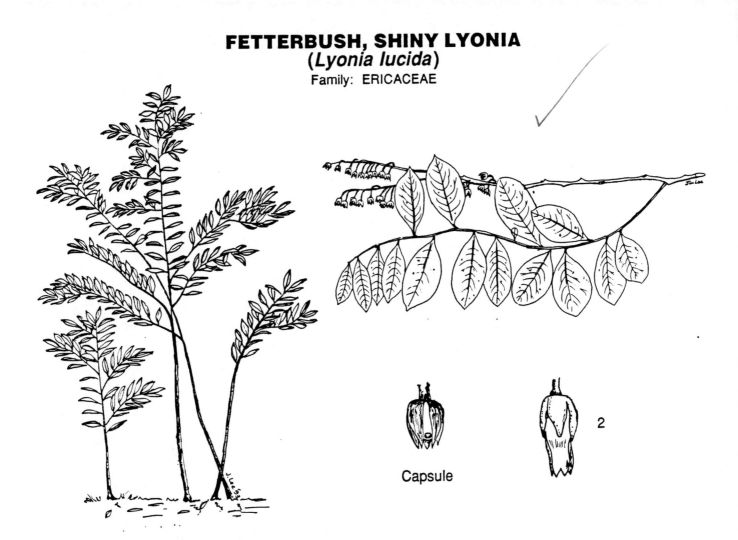

Capsule

2

IDENTIFYING CHARACTERISTICS (see below for details): 1) leaves — undersurface not rust-colored; distinct vein paralleling margin; 2) flowers — pink to red, rarely white, with persistent sepals.

DESCRIPTION	Medium-sized evergreen shrub with strongly angled branches.
GROWTH RATE	Moderate.
SIZE	Low-growing or up to 12' tall.
LEAVES	Alternate, shiny, elliptic to obovate; wide variation in leaf size (1¼–3½" long and ³/₈–1½" wide); smooth, rolled under margins with distinct vein paralleling the margin. Leaf tip pointed. Undersurface not rust-colored.
FLOWERS	Axillary clusters of urn-shaped pink to red, rarely white, blooms, ³/₈" long; corolla about twice as long as calyx and sepals are persistent. Winter and spring.
FRUIT	Ovoid, mainly glabrous capsules turning medium to dark brown at maturity.
BARK	Smooth upper stem; reddish.
RANGE	Found throughout the southeast; common in northern and central Florida.
HABITAT	Most commonly occurring in freshwater swamps and low pine flatwoods, often occurring as dense thickets. Sometimes on well-drained sands in scrub.
LANDSCAPE USAGE	Masses or specimen plant. Tolerant of a wide range of soil moisture conditions.
HUMAN EDIBILITY	None.
REMARKS	Wildlife food source.

SOUTHERN MAGNOLIA, BULL BAY
(*Magnolia grandiflora*)
Family: MAGNOLIACEAE

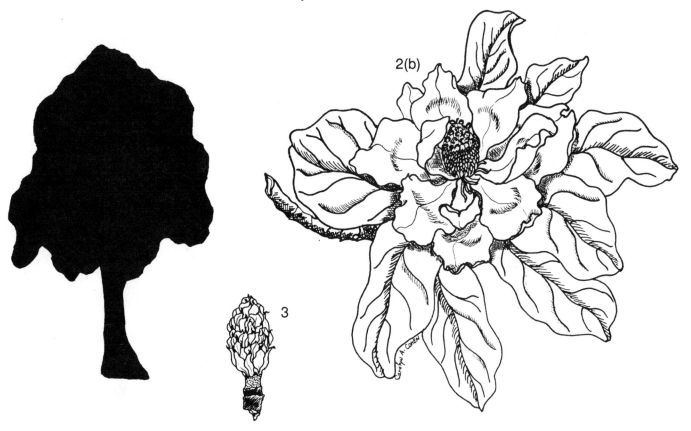

IDENTIFYING CHARACTERISTICS (see below for details): 1) Leaves — shiny dark green above, reddish-brown below; 2) flowers — (a) silky-hairy buds and (b) 6 to 8" in diameter; 3) fruit — pubescent, conelike, 3–5" long.

DESCRIPTION	Large, upright, spreading evergreen tree with a pyramidal shape.
GROWTH RATE	Rather slow.
SIZE	Generally 60–80', up to 100'; up to 4' in diameter.
LEAVES	Simple, alternate, large (5–8" long or longer and 2–4" wide), thick, leathery, elliptical or ovate, smooth margin; dark green and shiny above, rusty, sometimes silvery, beneath. Aromatic when bruised.
FLOWERS	Creamy white, very fragrant, 6–8" across with 6–12 petals and like-colored sepals; terminal, on stalks. Spring and summer.
FRUIT	Conelike, 3–5" long, yellowish to yellowish-brown pubescent, turn red to brown with age. Seeds are bright red and $^3/_4$" long, and hang from the "cone" on threads.
BARK	Gray to light brown; furrowed with age; rusty, feltlike when young.
RANGE	East North Carolina to DeSoto County in Florida, and west to eastern Texas.
HABITAT	Occurs naturally in rich hammock soils and on the border of river swamps.
LANDSCAPE USAGE	Widely cultivated for its ornamental value. Street, shade, specimen or framing tree.
HUMAN EDIBILITY	None.
REMARKS	Seeds eaten by birds and trees used for nesting and/or cover by wildlife. Selected cultivars are propagated from cuttings or by graftage. Leaves are used by florists. Used in woodworking and for firewood.

SWEET BAY MAGNOLIA
(*Magnolia virginiana*)
Family: MAGNOLIACEAE

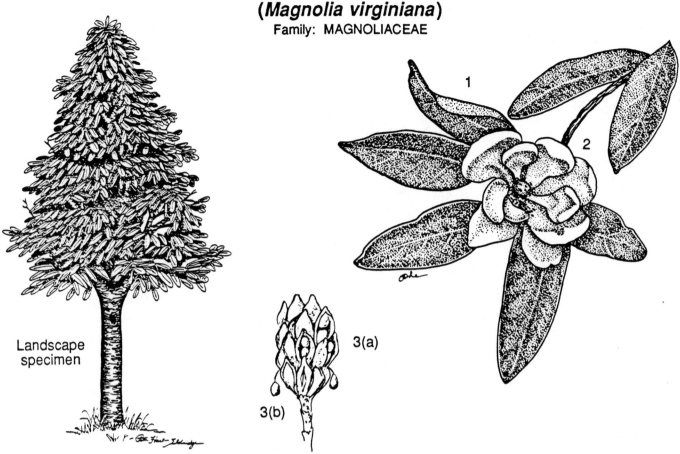

Landscape specimen

3(a)

3(b)

IDENTIFYING CHARACTERISTICS (see below for details): 1) Leaves — dull green above and silvery below; 2) flowers — 2 to 3" in diameter; 3) fruit — (a) glabrous, conelike, 1–2" long and (b) seeds on silky threads.

DESCRIPTION	Large evergreen tree distributed widely in bayheads, swamps, and along streams.
GROWTH RATE	Slow to moderate.
SIZE	Height generally 40–60', up to 90'; trunks to 3' diameter.
LEAVES	Simple, alternate, entire, thin, leathery and evergreen. Shiny green above and silvery beneath, 3–6" long, oblong to elliptic — narrowing at both ends. Aromatic when crushed.
FLOWERS	Creamy white, cup-shaped, very fragrant, 2–3" in diameter with 9–12 petals and like-colored sepals. Solitary and erect on ends of branches, blooms spring through fall.
FRUIT	Glabrous conelike fruit cluster, dark red, ovoid, 1–2" long; seeds red or brown, $\frac{1}{4}$" long, hang from "cone" on threads.
BARK	Smooth gray to pale brown trunk, broken into very small rectangular blocks by shallow cracks, often mottled by the presence of lichens.
RANGE	Coastal plain of eastern Massachusetts southward to Dade County in Florida, west to eastern Texas.
HABITAT	Low wet woodlands, wet flatwoods, hammocks.
LANDSCAPE USAGE	Specimen or street tree, and in naturalized landscapes. Attractive, fragrant flowers, shimmery silvery foliage and showy fruit. Tolerates shade, grows poorly in dry soil, but does well in wet areas.
HUMAN EDIBILITY	Leaves used as seasoning for fish and fowl.
REMARKS:	Fruit is eaten and tree is used for nesting and/or cover by wildlife. Beavers are often baited by fleshy roots of these trees, hence one of its common names, "beaver tree". Sprouts freely after fires, sometimes forms thickets. Timber has little economic importance.

WAX MYRTLE, SOUTHERN BAYBERRY
(*Myrica cerifera*)
Family: MYRICACEAE

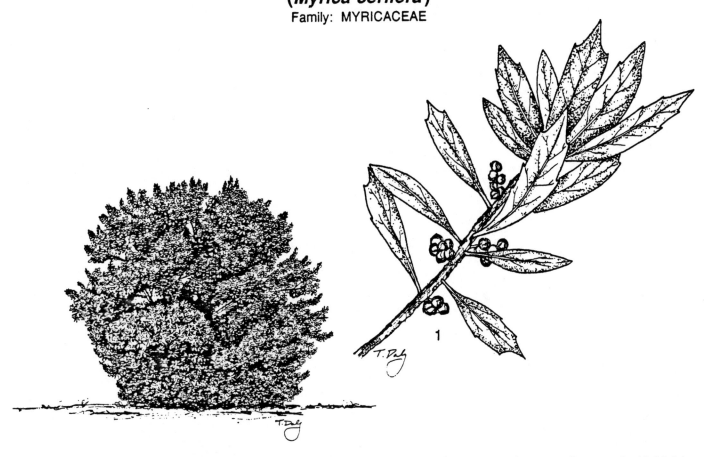

IDENTIFYING CHARACTERISTICS (see below for details): 1) fruit — persistent berries, heavily coated with bluish-white wax, in dense clusters along twigs; 2) leaves — lower surface somewhat hairy with orange glandular dots.

DESCRIPTION	Aromatic, evergreen shrub or small tree. Trunk is small and the slender upright, spreading branches form a round-topped head.
GROWTH RATE	Fast.
SIZE	Height up to 40' tall; trunk 8" in diameter.
LEAVES	Alternate, simple, 2–4" long, ¼–1" wide, coarsely toothed above the middle; smooth shining above, the lower surface somewhat hairy; orange dots (glands) on both surfaces; aromatic when crushed.
FLOWERS	Tiny flowers in short, scaly axillary catkins on branchlets of previous seasons. Dioecious.
FRUIT	Ripening in winter, on female plants, in dense clusters along twigs; persistent berries, ⅛" in diameter, heavily coated with bluish-white wax.
BARK	Silvery gray and smooth, young branchlets waxy and somewhat hairy.
RANGE	Found throughout the southeast, occurs in every county in Florida.
HABITAT	Common in moist, sandy soils of swamps, ponds, fields, ditch banks, forests or disturbed sites.
LANDSCAPE USAGE	Specimen or mass plantings for full sun or partial shade. Responds well to pruning. Benefits from being cut back to ground every 8–10 years. Dwarf selections may be available from some nurseries.
HUMAN EDIBILITY	Leaves were used for tea by Indians. Both fruits and leaves are used for flavoring.
REMARKS	Good wildlife food source. The wax on the berries can be used for making fragrant bayberry candles. Leaves are used as insect repellent. Root bark powder has been used as poultices and to alleviate nasal congestion and diarrhea.

SWAMP BLACK GUM, SWAMP TUPELO
(*Nyssa sylvatica* var. *biflora*)
Family: NYSSACEAE

IDENTIFYING CHARACTERISTICS (see below for details): 1) swollen base of trunk; 2) fruit — green, turning blue in fall; smooth, paired on stalks.

DESCRIPTION	A large deciduous tree with horizontal branches drooping gradually and gracefully at the ends.
GROWTH RATE	Slow to medium.
SIZE	Height to 70' (rarely to 100') with trunks 2½' in diameter, swollen at the base.
LEAVES	Alternate, simple, dark green above, paler below, ¾–3½" long, elliptic to lance-elliptic, tapered at base; mature leaves rarely toothed. Red in early fall, yellow or purple later.
FLOWERS	Inconspicuous, greenish, on long stalks at the base of new leaves. Male (in dense clusters) and female (few-flowered clusters) on separate trees; occasionally bisexual flowers occur. Spring.
FRUIT	Ovate, smooth, ½" long, usually paired on stalks; green, turning dark blue in fall.
BARK	Dark gray or brown, almost black with numerous deep interlacing furrows at maturity.
RANGE	Coastal plain of Delaware and Maryland south to the Caloosahatchee River in Florida, west to Louisiana.
HABITAT	Floodplain forests, wet flatwoods, cypress-gum ponds, pond and lake margins and shallow water of swamps.
LANDSCAPE USAGE	Excellent as a specimen tree or in naturalized areas. Will tolerate dry soils once established, but not high pH soils.
HUMAN EDIBILITY	None.
REMARKS	Fruits provide abundant food for birds and other animals and plant is used for nesting and/or cover by wildlife. In habitats subject to periodic burning, tree is smaller and scrubbier. One source of tupelo honey. Air pollution is the only serious problem for this tree.

WILD OLIVE, DEVILWOOD
(*Osmanthus americana*)
Family: OLEACEAE

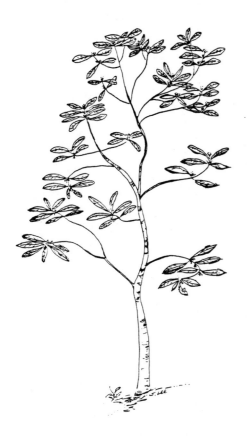

IDENTIFYING CHARACTERISTICS (see below for details): 1) fruit — dark bluish-purple ½" ovoid drupes on stalk; 2) flowers — small, fragrant, funnel-shaped.

DESCRIPTION	Evergreen shrub or small tree with loose, open habit, found in a wide range of wooded habitats.
GROWTH RATE	Slow.
SIZE	Height 15–25', up to 45'; trunk diameter up to 1'.
LEAVES	Opposite, simple, leathery, shiny, variable in shape — elliptic, oblong elliptic, oblanceolate — usually 4–5", but up to 9" long, 2" wide, smooth, curled under margins. Upper surface dark green and shiny, lower surface paler and duller.
FLOWERS	Small, fragrant greenish to creamy white funnel-shaped, 4-lobed flowers in small dense clusters in the axils of the leaves of previous year's growth; open in early spring.
FRUIT	Oval, bitter drupe, ¼–½" in diameter, ⅜–¾" long, 1-seeded, dark bluish-purple; on stalks; persists into spring.
BARK	Pale.
RANGE	Coastal plain of southeast Virginia to central Florida, west to Louisiana.
HABITAT	Wooded sites, well-drained uplands, bottomlands and flatwoods with periodic inundation.
LANDSCAPE USAGE	Naturalized landscapes or foundation plant in formal setting. Very fragrant so large plantings could be overpoweringly fragrant.
HUMAN EDIBILITY	None.
REMARKS	Fruits provide abundant food for birds. Genus name from Greek *osme* (odor) and *anthos* (flower).

REDBAY
(*Persea borbonia*)
Family: LAURACEAE

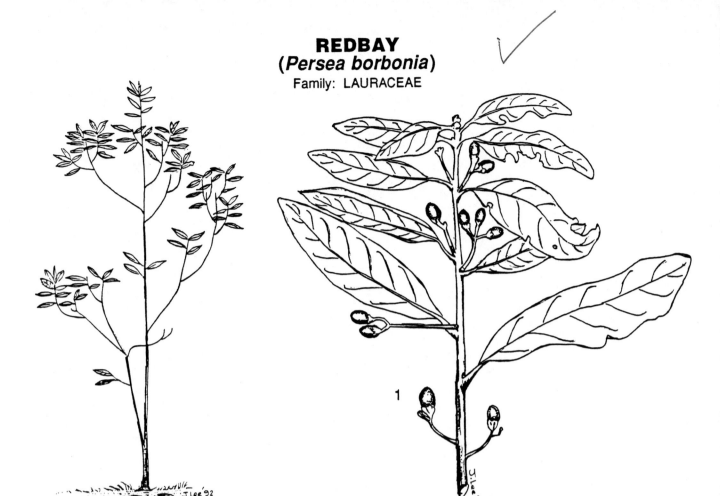

IDENTIFYING CHARACTERISTICS (see below for details): 1) fruit — dark bluish-black $^3/_8$–$^1/_2$" drupes on $^1/_2$–1" orange stalk; 2) leaves — aromatic, sparsely pubescent on lower surface.

DESCRIPTION	Large shrub to medium-sized tree, evergreen and aromatic.
GROWTH RATE	Medium.
SIZE	Six to 70' tall; trunk up to 3' in diameter.
LEAVES	Alternate, entire, elliptic, 2–8" long, $^3/_4$–1$^1/_2$" wide, dark green and shiny above and lighter green smooth to very sparsely pubescent beneath.
FLOWERS	Small, pale yellow, compact axillary cymes on hairy stalks; occur in spring.
FRUIT	Dark bluish-black drupes, imperfectly spherical, $^3/_8$–$^1/_2$" in diameter on $^1/_2$–1" orange peduncle; maturing in late summer to fall.
BARK	Mature bark brownish-gray, very rough with approximately vertical, interlacing fissures between ridges. Twigs green to reddish, sparsely pubescent.
RANGE	Delaware south to south Florida and west to Texas.
HABITAT	Various mesic to dry sites; typically in drier and better drained soils than *Persea palustris* (swampbay).
LANDSCAPE USAGE	Naturalized landscapes, specimen and/or shade plant.
HUMAN EDIBILITY	Leaves used for flavoring.
REMARKS	Wildlife food source. Dense orange-colored hardwood takes a high polish and is used in woodworking. Distinguishable from swampbay (*P. palustris*) by the absence of a hairy leaf surface and by differences in habitat preference as mentioned above. A fairly reliable field character for members of this genus is the presence of "warty" or gall-like growths along portions of the leaf margins.

SWAMPBAY
(*Persea palustris*)
Family: LAURACEAE

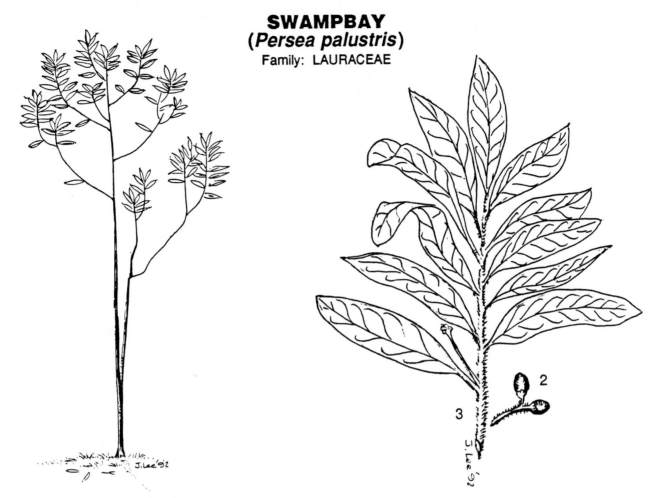

IDENTIFYING CHARACTERISTICS (see below for details): 1) leaves — aromatic and lower surface rusty, with densely matted hairs; 2) fruit — small, bluish-black drupes on 1–2" long peduncle; 3) bark — hairy young twigs.

DESCRIPTION	Evergreen shrub or small tree.
GROWTH RATE	Slow.
SIZE	Up to 40' in height.
LEAVES	Alternate, entire, broadly elliptic, 1½–4¾" long, 1–2" wide, smooth to slightly pubescent above and densely hairy, white to rusty-brown below. Aromatic when crushed. Leaf stalk densely hairy.
FLOWERS	Small, cream-colored on densely pubescent axillary inflorescences. Spring and early summer.
FRUIT	Ellipsoid to globose bluish-black drupes approximately ⅜" long on 2–3" hairy peduncles.
BARK	Tightly furrowed. Young twigs are densely hairy.
RANGE	Coastal plain, Virginia to south Florida, westward to southeast Texas.
HABITAT	Cypress and hardwood swamps with periodic inundation, as well as soils with high water table conditions (poorly drained).
LANDSCAPE USAGE	Should be incorporated into landscapes where they occur naturally; however, can tolerate moderately dry conditions.
HUMAN EDIBILITY	Leaves used for flavoring/seasoning.
REMARKS	Wildlife food source. Distinguishable from redbay (*P. borbonia*) by the presence of a hairy leaf surface and by differences in habitat preference as mentioned above. A fairly reliable field character for members of this genus is the presence of "warty" or gall-like growths along portions of the leaf margins. Larger specimens probably difficult to transplant.

SAND PINE
(*Pinus clausa*)
Family: PINACEAE

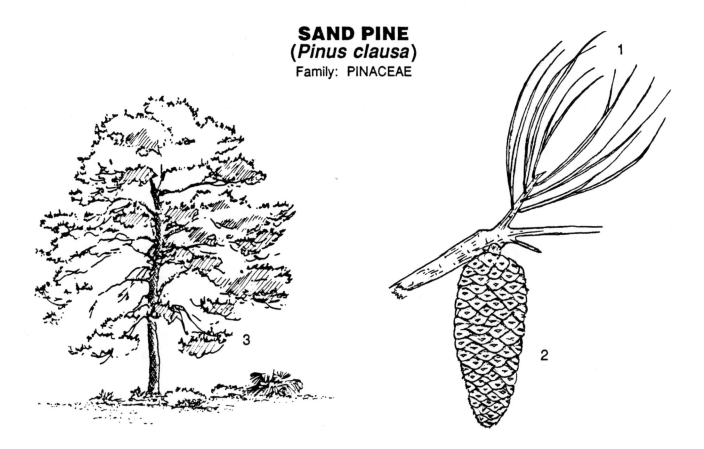

IDENTIFYING CHARACTERISTICS (see below for details): 1) leaves — short (3") needles occurring in 2s; 2) fruit — 2–3" long cones; 3) lower branches persist on trunk.

DESCRIPTION	Small to medium-sized evergreen tree with much-branched smooth twigs. Conical, with lower lateral branches persistent on trunk.
GROWTH RATE	Fast.
SIZE	Twenty to 30', up to 70'; will reach 2' in diameter.
LEAVES	Needles in 2s, often twisted, about 3" long, short compared to other southern pines. Fascicle sheaths $\frac{1}{8}$–$\frac{3}{16}$" long.
FLOWERS	Not applicable. The gymnosperms bear naked seeds, usually on the scale of a cone (see next item).
FRUIT	$\frac{3}{4}$" winged seeds in 2–$3\frac{1}{2}$" long cones, maturing after 2 years. Persist, open or unopened, on trees for many years.
BARK	Tan or brownish, smooth thin bark on smaller trunks and branches; reddish-brown and scaly when larger.
RANGE	Extreme southern Alabama and in Florida on coastal sand dunes as far south as Dade and Lee counties, as well as the scrubs of the interior part of the state into north Florida.
HABITAT	Commonly found on deep, generally poor, but well-drained soils in association with evergreen scrub oaks; will not tolerate high water tables or fire.
LANDSCAPE USAGE	Good pine for scrub or water-conserving landscaping; established trees may be adversely affected by subsequent irrigation. Sensitive to construction damage and will not transplant well.
HUMAN EDIBILITY	Raw or roasted seeds can be eaten; young needles made into tea; inner bark (cambium layer) can be dried and ground into flour.
REMARKS	Used by wildlife for food, nesting materials, nest sites and/or cover. Two geographical varieties: the Ocala and the Choctawhatchee. Sand pines, especially the Ocala variety, have the ability to reseed burned-over areas. Their serotinous cones (cones that remain closed long after the seeds inside are ripe) require heat (fire or intense sunlight) to open and release seeds. The Choctawhatchee is preferred for Christmas trees. Sand pines are used for pulpwood.

SLASH PINE
(*Pinus elliottii*)
Family: PINACEAE

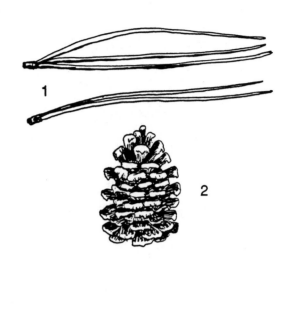

IDENTIFYING CHARACTERISTICS (see below for details): 1) leaves — 7–12" long in clusters of 2s or 3s; 2) fruit — borne in cones that are 3–6" long.

DESCRIPTION	Tall, straight pine with scaly, rough twigs and lush, dark green foliage. Crown composed of short, stout branches. Rusty-silver terminal buds.
GROWTH RATE	Rapid.
SIZE	To 100' tall; trunk 2–3' in diameter.
LEAVES	Persistent needles in 2s and 3s, 7–12" long. Fascicle sheaths are $^3/_8$–$^5/_8$" long.
FLOWERS	Not applicable. The gymnosperms bear naked seeds, usually on the scale of a cone (see next item).
FRUIT	Stalked, oval cones, 3–6" long fall from trees in 2nd year. Winged seeds fall in autumn of 2nd year.
BARK	Dark gray, deeply furrowed, breaking into irregular rectangular plates, twice as long as wide, exposing reddish-brown inner bark.
RANGE	Coastal plain from South Carolina to Highlands County, FL and west to southeast Louisiana. Its range is expanding.
HABITAT	Found primarily along areas of inland flatwoods. Prefers moist soils, but will tolerate a wide range of soil types, including well-drained upland soils.
LANDSCAPE USAGE	Good landscape tree, best grown in clusters.
HUMAN EDIBILITY	Seeds can be eaten raw or roasted; young needles made into tea; inner bark (cambium layer) can be dried and ground into flour.
REMARKS	Eagles use their crown for nests and red-cockaded woodpeckers excavate nest cavities in the trunks of older trees infected with red heart fungus. Used by wildlife for food and nesting materials as well as nesting sites and cover. Wood is used for lumber and pulpwood; turpentine is collected from the tree; and cones and needles are used for handicrafts.

LONGLEAF PINE
(*Pinus palustris*)
Family: PINACEAE

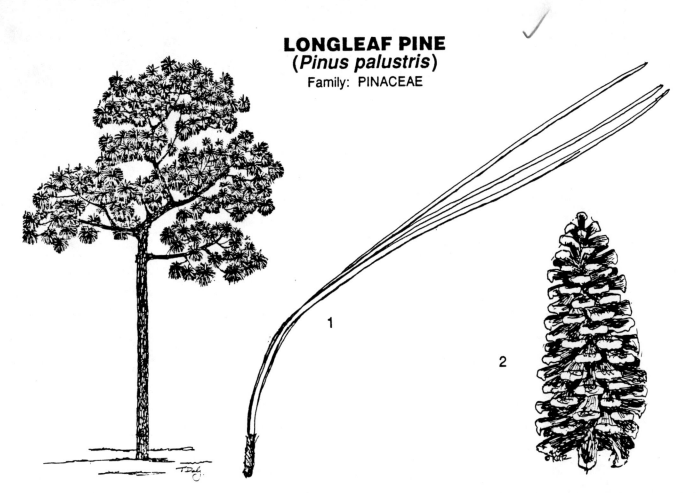

IDENTIFYING CHARACTERISTICS (see below for details): 1) leaves — 8–18" long needles in clusters of 3s; 2) fruit borne in 5–10" long cones; 3) silvery-white terminal buds.

DESCRIPTION	Large tree with a long, bare trunk and small open crown. Silvery-white terminal buds.
GROWTH RATE	Moderate.
SIZE	Eighty to 120' tall; trunk 2–2½' in diameter. Twigs very stout, ½" or greater in diameter.
LEAVES	Needles in 3s, 8–18" long, fan out at end of branches. Fascicle sheaths usually greater than ½" long.
FLOWERS	Not applicable. The gymnosperms bear naked seeds, usually on the scale of a cone (see next item).
FRUIT	Long cones — 5–10" — fall in autumn of second year. Seeds are winged.
BARK	Dark reddish-brown, furrowed, breaking into large irregular plates exposing reddish-brown inner bark. Small trunks, gray and rough. Twigs are dark brown and terminate with large white buds.
RANGE	Southeast U.S. from Virginia to Florida and west to Texas.
HABITAT	Found on a variety of soils. Flatwoods and with deciduous scrub oaks on sand ridges.
LANDSCAPE USAGE	Good landscape tree especially for use in dry upland habitats where it was formerly abundant.
HUMAN EDIBILITY	Seeds eaten raw or roasted; young needles made into tea; inner bark (cambium layer) can be dried and ground into flour.
REMARKS	Used by wildlife for food, cover, nesting materials and/or nest sites. The original source of turpentine and resinous products (naval stores); it makes good lumber for construction, often after tapping for turpentine and resin. One of the most distinctive and important of the southern conifers. Adapted to habitats where fire is a frequent occurrence — as a seedling will stay in a grass stage for first 3–6 years, then shoots up rapidly and may then attain enough height to survive subsequent fires. Although the longleaf pines are slow starters they will catch up with slash pines in height by age 15. This is the pine most reduced in acreage by human activities; endangered and should be planted more than it is. Now being recommended for lumber plantations where appropriate.

POND PINE
(*Pinus serotina*)
Family: PINACEAE

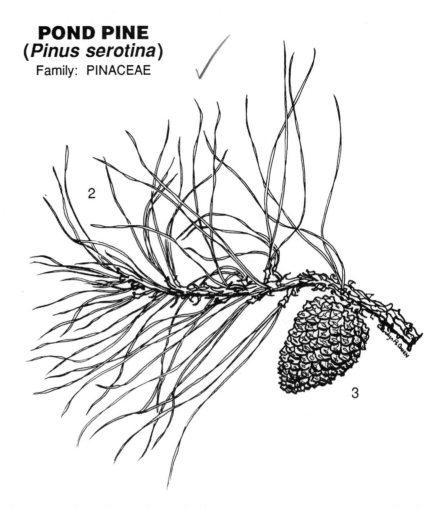

IDENTIFYING CHARACTERISTICS (see below for details): 1) tufts of twigs and needles grow on main trunk; 2) leaves — 6–8" long needles in clusters of 3s, rarely 4s; 3) fruit — 2–2½"-long cones.

DESCRIPTION	A small to medium-sized tree found in pure stands on wet sites. Crown branches are irregular, crooked and gnarly; twigs slender and rough-scaly.
GROWTH RATE	Slow to medium.
SIZE	Average height 40–50', but up to 70'; trunk diameter 1–2'.
LEAVES	Persistent needles in 3s, rarely in 4s, generally 5–8" long. Fascicle sheaths are ¼–⅜" long.
FLOWERS	Not applicable. The gymnosperms bear naked seeds, usually on the scale of a cone (see next item).
FRUIT	Stalkless in appearance, round to oval-shaped cones are 2–3" long, very broad when open with a sharp, minute spine on each scale face. Cones are persistent on trees for many years. Small winged seeds drop during fall of second year.
BARK	Dark gray or reddish-brown; coarse, furrowed, flaking into narrow vertical plates exposing dark-brown inner bark.
RANGE	Found over the coastal plain from southern New Jersey (sparse) to central and northwestern Florida and adjacent Alabama.
HABITAT	Commonly found on poorly drained soils in low flatwoods and near flatwoods ponds.
LANDSCAPE USAGE	Other pines are better suited for landscape planting.
HUMAN EDIBILITY	Seeds can be eaten raw or roasted; young needles made into tea; inner bark (cambium layer) can be dried and ground into flour.
REMARKS	Food, cover, nesting materials and/or nest site source for wildlife. Tufts of twigs and needles growing on the main trunk are good identification traits. Increasing in importance as a pulpwood source.

TURKEY OAK
(*Quercus laevis*)
Family: FAGACEAE

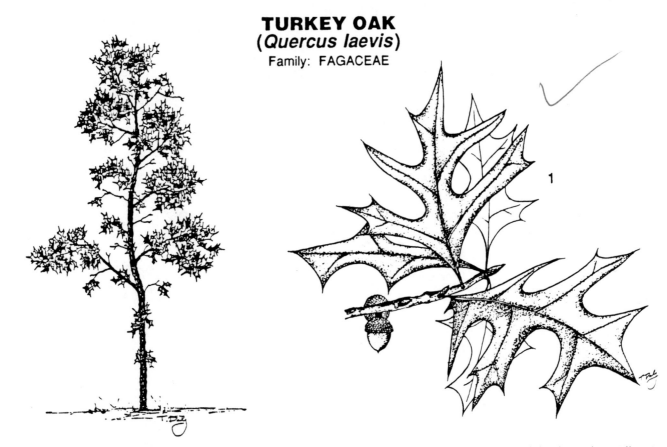

IDENTIFYING CHARACTERISTICS (see below for details): 1) leaves — deciduous, deeply lobed, turning yellow to red in fall; 2) bark — deep-furrowed, gray.

DESCRIPTION	A deciduous tree with an irregular, rounded crown.
GROWTH RATE	Medium to fast.
SIZE	Commonly 20–30', no more than 60' tall; trunk 1–2' diameter. Reaches a good size, but of short stature on dry sandy ridges and hammocks.
LEAVES	Alternate, simple, 3–12" long, 3–6" wide with 3–7 narrow, curved and often deeply divided and toothed lobes, prickles on tips of leaves; heavily veined, bright yellow-green and lustrous above, paler underside and somewhat hairy, but only in vein axils. Leaves turn yellow to red in fall.
FLOWERS	Appearing with the leaves; unisexual — the staminate ones are in slender, 4–8" hairy catkins and the pistillate are on short, stout, hairy axillary stalks.
FRUIT	Acorn: dull, light brown, $^3/_4$" broad, 1" long, bow-shaped, sessile or with a short stalk; hairy (inside) cup encloses $^1/_3$–$^1/_2$ of the nut. One to 3 acorns occur together. Takes two years to make acorns.
BARK	Gray to black, scaly-ridged.
RANGE	From southeast Virginia to central Florida and west to Louisiana.
HABITAT	Well-drained sandy ridges. Commonly associated with longleaf pine, bluejack oak and dwarf post oak. A drought resistant tree.
LANDSCAPE USAGE	An excellent source for fall color, but established trees will not tolerate excessive amounts of fertilizer or water.
HUMAN EDIBILITY	Acorn meat is yellow to orange and very bitter ("red-black" oak group); edible if tannin is leached out, but not worth time and effort when "white" oak group acorns are available.
REMARKS	Fruit and insects harvested from trees are used as food and the tree is used for nesting and/or cover by wildlife. Cut for firewood and leaves with fall color are used as decorations. Reproduction, especially after burns, may be from underground runners. Narrow 3-lobed leaves sometimes resemble a turkey's footprint. Laevis (Latin) means smooth and refers to hairless leaves.

DIAMOND-LEAF OAK, LAUREL OAK
(*Quercus laurifolia*)
Family: FAGACEAE

IDENTIFYING CHARACTERISTICS (see below for details): 1) fruit — acorns, solitary or in pairs, with reddish, pubescent scales on cup. 2) leaves — yellow midrib.

DESCRIPTION	A late winter/early spring deciduous tree with a broad, round-topped, dense crown.
GROWTH RATE	Rapid.
SIZE	Fifty to 60' tall, occasionally 100'; trunk 2–4' in diameter.
LEAVES	Alternate, simple, semi-evergreen, elliptical or rarely oblong-obovate, 2–5$\frac{1}{2}$" long, $\frac{1}{2}$–2" wide; apex is acute, base is wedge-shaped; margin is entire or rarely irregularly lobed; shiny green above, pale below with yellow midrib; petioles are stout, yellow, and $\frac{1}{4}$" long. Seedling and root sucker leaves are generally several-lobed.
FLOWERS	Small, unisexual; male flowers in 3" catkins; female flowers are on short stalks in leaf axils.
FRUIT	Acorn: solitary or in pairs, commonly subsessile; nut, ovoid or hemispherical, brownish-black, $\frac{1}{2}$" long; cup is thin, saucer-shaped, with reddish, pubescent scales and covers about $\frac{1}{4}$ of nut.
BARK	Up to $\frac{1}{2}$" thick, dark reddish-brown; at first smooth, becoming grayish, then darker and divided into deep fissures separated by broad, flat ridges.
RANGE	Southeast Virginia south to Florida and west to Texas and Arkansas.
HABITAT	Moist to wet well-drained sandy soils near streams and swamps.
LANDSCAPE USAGE	A common ornamental in central Florida, it grows on a variety of soil types. Easy to maintain shade and street tree.
HUMAN EDIBILITY	Acorn meat is yellow to orange and very bitter ("red-black" oak group); edible if tannin is leached out, but not worth time and effort when "white" oak group acorns are available.
REMARKS	Fruit and insects harvested from trees are used as food and the tree is used for nesting and/or cover by wildlife. Used locally as fuel. Longevity usually 30–50 years. Semi-evergreen refers to fact that as old leaves drop, new leaves are already developing. Some separate these into distinct species, but Wunderlin (see bibliography) considers it all one variable species.

MYRTLE OAK
(*Quercus myrtifolia*)
Family: FAGACEAE

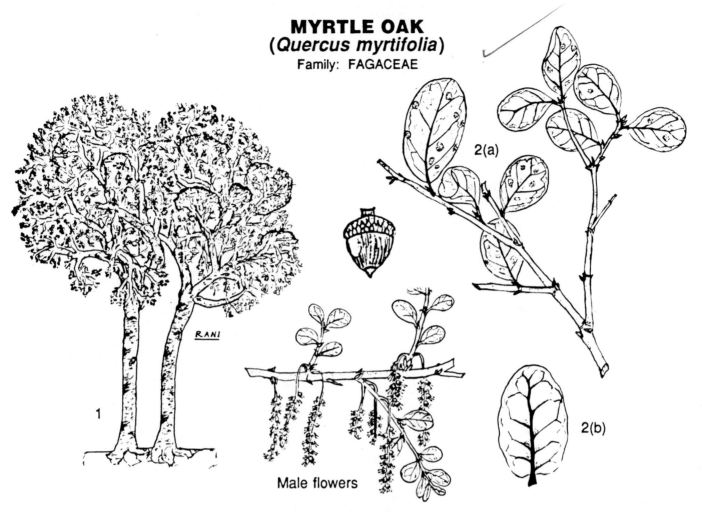

2(a)

2(b)

1

RANI

Male flowers

IDENTIFYING CHARACTERISTICS (see below for details): 1) small, thicket-forming; 2) leaves — (a) smooth above, slightly revolute, widest above the middle, and (b) underside may have sparse hairs or small scales when mature.

DESCRIPTION	A small, evergreen, thicket-forming shrub or small tree.
GROWTH RATE	Slow.
SIZE	Seldom over 35' in height; trunk 4–8" in diameter.
LEAVES	Dark green, leathery, $\frac{1}{2}$–2" long, $\frac{1}{3}$–1" wide, ovate to oblong-ovate, widest above the middle, tips sharp or rounded; margins smooth and turned under; smooth on upper surface, lower surface may have sparse hairs (usually at vein angles - see underside leaf drawing above) or small golden scales when mature.
FLOWERS	Small, unisexual; male flowers in catkins. Both male and female flowers on the same tree.
FRUIT	Acorn: dark lustrous brown, nearly round striated nuts, about $\frac{1}{2}$" in diameter; saucerlike to top-shaped cup covers $\frac{1}{3}$ of the nut.
BARK	Light gray, smooth but becoming slightly furrowed.
RANGE	South Carolina to Mississippi inclusive of peninsular Florida.
HABITAT	Oak scrub, pine flatwoods, well-drained, white sandy ridges in nearly impenetrable thickets.
LANDSCAPE USAGE	Suitable for naturalized landscapes (see Remarks).
HUMAN EDIBILITY	Acorn meat is yellow to orange and very bitter ("red-black" oak group); edible if tannin is leached out, but not worth time and effort when "white" oak group acorns are available.
REMARKS	Fruit and insects harvested from trees are used as food and the tree is used for nesting and/or cover by wildlife. Short-lived oak. This tree has little commercial value.

WATER OAK
(*Quercus nigra*)
Family: FAGACEAE

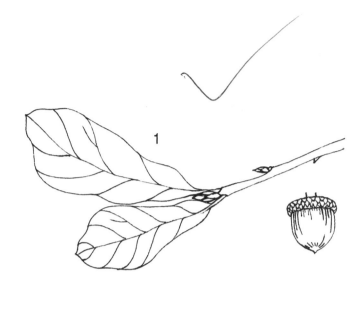

IDENTIFYING CHARACTERISTICS (see below for details): 1) leaves — variably shaped, generally tip is wide and base is wedge-shaped; 2) usually in wetter locations than most other oaks.

DESCRIPTION	A tall, slender tree with a round-topped, symmetrical crown of ascending branches.
GROWTH RATE	Fast in its early years.
SIZE	50–95' in height; trunk 2–3½' in diameter.
LEAVES	Alternate, simple, deciduous, variable in shape, but mostly spatulate, oblong-obovate, or oblong-lanceolate; 2–4" long, 1–2" wide (occasionally 6" long); apex is acute to broadly obtuse, base is usually wedge-shaped and margin is variable — (1) entire, (2) 3-lobed at the apex, or (3) variously lobed, the latter form being largely restricted to vigorous sprouts and juvenile plants; dull, bluish-green above, paler below, often with rusty axillary tufts. Petioles are short, stout and flattened.
FLOWERS	Small, unisexual; male flowers in catkins. Female flowers on same tree as male.
FRUIT	Acorn: solitary or occasionally in pairs, often short-stalked; nut is ovoid to hemispherical, light brown to nearly jet-black with a pubescent apex; about ½" long; thin, saucer-shaped, reddish-brown, pubescent cup covers about ¼ of the nut.
BARK	At first smooth and brown, becoming gray-black with shallow, rough, scaly ridges.
RANGE	Southern New Jersey to central Florida west to eastern Texas and north to southeast Missouri.
HABITAT	Typically a bottomland species, but seldom occurring in permanent swamps; usually associated with other hardwoods and, under favorable conditions, the most abundant species in the stand.
LANDSCAPE USAGE	Because of its fast growth during the first years, ease of propagation and pleasing form, it is a favorite street and lawn tree, but it must have adequate space.
HUMAN EDIBILITY	Acorn meat is yellow to orange and very bitter ("red-black" oak group); edible if tannin is leached out, but not worth time and effort when "white" oak group acorns are available.
REMARKS	Fruit and insects harvested from trees are used as food and the tree is used for nesting and/or cover by wildlife. Used as fuel and timber.

LIVE OAK
(*Quercus virginiana*)
Family: FAGACEAE

IDENTIFYING CHARACTERISTICS (see below for details): 1) wide spread branches, often dipping close to ground; 2) fruit — acorns are long-stalked and occur singly or in clusters of 3–5.

DESCRIPTION	Evergreen tree with massive wide-spread limbs branching close to the ground and forming a broad, low, dense, round-topped crown.
GROWTH RATE	Moderate; responds well to fertilizer and irrigation when young.
SIZE	40–50' in height; trunk 3–4' in diameter; branches spread to 100' or more.
LEAVES	Leathery, alternate, simple, falling in the spring after the new foliage has appeared; $1/2$–4" long, $3/8$–$2^1/2$" wide; oblong-obovate, or elliptical; apex is obtuse, base is acutely wedge-shaped and margin is slightly revolute and entire or, rarely, toothed; dark, lustrous green above, usually pale-pubescent below; petioles are stout and $1/4$" long.
FLOWERS	Small, unisexual; male flowers in catkins. Monoecious.
FRUIT	Acorns: set yearly, long-stalked, single or clusters of 3–5 nuts; nut is ellipsoid or obovoid, brownish-black, $3/4$–1" long; top-shaped cup — scaly outside, smooth inside — encloses $1/3$ of the nut.
BARK	Dark red-brown to grayish-brown, up to 1" thick, furrowed, separating into small, appressed scales and becoming deeply fissured with age.
RANGE	Virginia south to Florida, west to west-central Texas. Also found in Mexico and Cuba.
HABITAT	Tolerant of many soil types, except wet. Hammocks, salt marsh borders, flatwoods, open roadsides, open upland woods, margins of depressions.
LANDSCAPE USAGE	Shade tree for large yards. Good tree for parks and street plantings. Very tolerant of urban conditions.
HUMAN EDIBILITY	Acorn meats are cream-colored and sweet ("white" oak group); acorn meat can be ground into flour and used as any other flour for baking. Also can be made into a delicious candy (see bibliography — Deuerling and Lantz).
REMARKS	Fruit and insects harvested from trees are used as food and the tree is used for nesting and/or cover by wildlife. Most hurricane-resistant trees available. Recognized as the tree of the deep south. Used as firewood; historically used for ship's ribs and knees, wheel hubs, and wooden cogs. Branches often have Spanish Moss hanging from them.

SHINING OR WINGED SUMAC
(*Rhus copallina*)
Family: ANACARDIACEAE

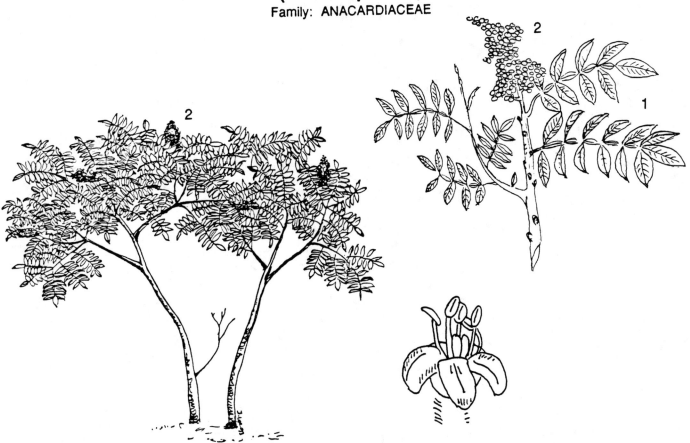

IDENTIFYING CHARACTERISTICS (see below for details): 1) leaves — rachis of compound leaf is winged; foliage turns red in autumn; 2) fruit — bright red, occurring in terminal clusters which persist into winter.

DESCRIPTION	Shrub or small, deciduous tree, often occurring in thickets.
GROWTH RATE	Moderate to fast.
SIZE	Up to 25–30' in height, usually smaller; trunk up to 8–10" in diameter.
LEAVES	Alternate, odd-pinnate up to 12" long; 7–21 leaflets, upper surface is shiny, pale and downy beneath; rachis is winged. Turn red in autumn.
FLOWERS	Small ($\frac{1}{8}$" wide), greenish cream, in 4–8" terminal panicles; occur in summer.
FRUIT	Downy, bright red, ovate, $\frac{1}{8}$" in diameter; fruiting clusters mature in fall and persist into the winter.
BARK	Twigs thin, reddish-brown; pubescent with conspicuous, wartlike reddish lenticels; sap is watery.
RANGE	Everywhere in Florida north of the Florida Keys. Mostly east of the Mississippi River in the U.S.
HABITAT	Poor soils of low moisture. Dry roadsides, old fields, clearings, waste places, open uplands and shady hammocks.
LANDSCAPE USAGE	Planted as an ornamental specimen for its shiny green leaves that turn red in the fall and its showy fruit. Used for naturalistic plantings. Good where there is plenty of space because of its coarse-growing, suckering growth habit. Plant in sunny situations on well-drained soils.
HUMAN EDIBILITY	Fruit contains a covering of malic acid; can be made into a lemonade-like drink. Also makes a delicious jelly.
REMARKS	Wildlife eat the fruit and twigs. Sumac thickets provide cover for birds and other wildlife. This species of sumac is non-poisonous. Indians used leaves as substitute for tobacco.

CABBAGE PALM, CABBAGE PALMETTO
(*Sabal palmetto*)
Family: ARECACEAE or PALMAE

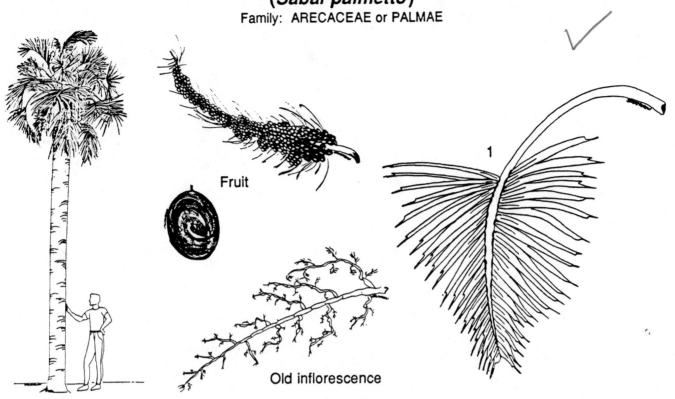

Fruit

Old inflorescence

1

IDENTIFYING CHARACTERISTICS (see below for details): 1) leaves — smooth-stalked, fanlike with downward curving midrib; 2) trunk may be criss-crossed with boots or smooth.

DESCRIPTION	A medium-sized, spineless, evergreen palm tree with very large, fan-shaped leaves spreading around the top.
GROWTH RATE	Very slow.
SIZE	30–80' in height; trunk up to 2' in diameter.
LEAVES	Alternate, smooth-stalked, fan-shaped, firm, 4–8' long, midrib curving down; variable in color (green to blue), margins deeply divided, bearing numerous threadlike fibers.
FLOWERS	Small ($\frac{1}{4}$" in diameter), white, fragrant, blossoms in drooping clusters, 2–2$\frac{1}{2}$' long; present in early summer.
FRUIT	Round, $\frac{1}{3}$" in diameter, blackish (when ripe), maturing in autumn; 1-seeded.
BARK	Gray-brown; rough or shallow-ridged. May or may not be covered with old leaf bases.
RANGE	Peninsular Florida, Gulf coastal areas in Florida and near coast in Georgia and the Carolinas.
HABITAT	Dunes (coastal and inland), wet forests, flatwoods, and seasonally wet prairies.
LANDSCAPE USAGE	Street, accent, framing and specimen tree. Best effect when planted in clusters. Will tolerate quite dry soil conditions once established.
HUMAN EDIBILITY	Fruits are eaten raw and made into syrup and jelly; the pulp is very sweet and prunelike in flavor. Dried fruits are made into a meal used to make bread. The bud (heart of palm) is edible, but removal of the bud kills the tree and is therefore not encouraged; the bud of saw palmetto (*Serenoa repens*), while smaller, is just as good or better than that of *Sabal palmetto* and harvesting it does not kill the plant.
REMARKS	The cabbage palm is the official tree of the State of Florida and is the most common palm tree found in the state. The tree is very versatile: trunks are used for wharf pilings, docks and poles; leaves are made into baskets, hats and mats; fibers of young leaves are fashioned into brushes. In addition to the human uses listed, this plant is also a source for food and nesting and/or cover for wildlife.

COASTAL PLAIN WILLOW
(*Salix caroliniana*)
Family: SALICACEAE

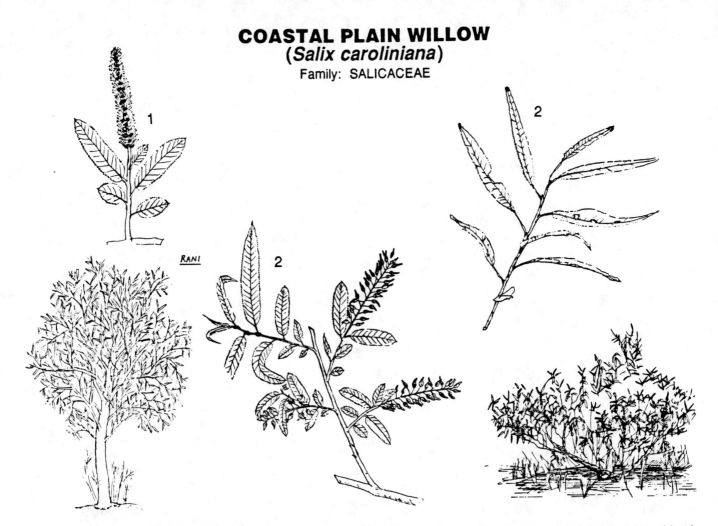

RANI

IDENTIFYING CHARACTERISTICS (see below for details): 1) flowers — 3–4" long yellowish to greenish catkins in spring; 2) leaves — finely-toothed margins commonly tipped with yellow or red glands.

DESCRIPTION	Shrub or small deciduous tree with multiple spreading and/or slightly drooping branches.
GROWTH RATE	Rapid.
SIZE	Height 30'; trunk about 1' in diameter.
LEAVES	Stalked, alternate, simple, thin, green and shining above, pale, silvery and glabrous (sometimes glaucous) beneath; 2 to 8" long and $\frac{1}{2}$–$\frac{3}{4}$" wide, lance-shaped and margins finely toothed; teeth commonly tipped with yellowish, sometimes reddish, glands.
FLOWERS	Catkins, greenish or yellowish, 3–4" long, at the ends of newly leafing twigs in spring.
FRUIT	Pointed capsules containing many cottony white hairy seeds, $\frac{1}{4}$" long, maturing in late spring or early summer before leaves fully mature.
BARK	Reddish twigs. Gray to black, fairly smooth, furrowed into broad scaly ridges on older growth.
RANGE	All of Florida, most of southeastern U.S. from southern Pennsylvania to central Texas.
HABITAT	Swamps, marshy places, and along streams, rivers, ponds and lake margins.
LANDSCAPE USAGE	Wet locations. Efficient for erosion control along streams.
HUMAN EDIBILITY	None.
REMARKS	Similar to the Florida willow (*Salix floridana*) which has elliptic to ovate-oblong leaves which are 2" wide and soft-pubescent below and the black willow (*Salix nigra*) which has leaves green on both sides. Roots and buds have been used as medicinal substitute for aspirin.

AMERICAN ELDER, ELDERBERRY
(*Sambucus canadensis*)
Family: CAPRIFOLIACEAE

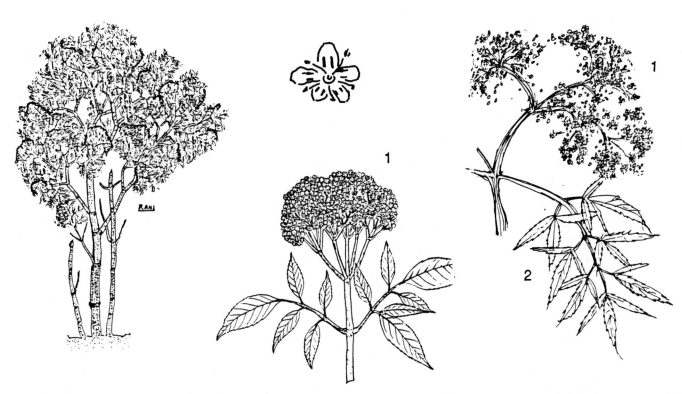

RANI

1

1

2

IDENTIFYING CHARACTERISTICS (see below for details): 1) flowers — bloom year round in Florida in flat-topped clusters on long stalks; 2) leaves — pinnately compound, 5–11 leaflets with sharply toothed margins; 3) prominent lenticels and numerous vertical furrows on older stems.

DESCRIPTION	Large weedy shrub or small tree common to roadside ditches and other unshaded moist soil sites.
GROWTH RATE	Rapid.
SIZE	Height to 20'; trunk to 6" in diameter.
LEAVES	Opposite, persistent, pinnately compound, 4–9" long with 5 to 11 smooth dark green leaflets, each $1\frac{1}{2}$–7" long. Leaflets elliptic with pointed tips and wedge-shaped bases, margins sharply toothed. Lower leaflets may be 2- or 3-lobed. Crushed leaves have rank odor.
FLOWERS	Very small, fragrant, 5-lobed white flowers bloom year round (in Florida) in flat-topped, 4–10" clusters (cymes) on long stalks.
FRUIT	Dark purplish-black, smooth, globose, $\frac{1}{4}$" in diameter.
BARK	Grayish-brown with prominent lenticels and numerous vertical furrows on older stems.
RANGE	Eastern U.S., including all of Florida.
HABITAT	Common in moist to wet open places — forest edges, swamps, springs and disturbed areas.
LANDSCAPE USAGE	Can be used as a screen in full-sun moist to wet locations, requires heavy pruning. Easily established and cutting to ground every few years will rejuvenate this plant.
HUMAN EDIBILITY	Berries are used for pies, cake, jelly and wine; blossoms are used to make tea and champagne, and are added to pancakes for flavor and lightness. Also flowerheads can be dipped in batter and deep fried and eaten out-of-hand using the stem as the handle.
REMARKS	Berries are eaten by birds and other wildlife. Although toys, such as whistles, have been made from the stems in the past, doing so may make the whistle-user sick.

SAW PALMETTO
(*Serenoa repens*)
Family: ARECACEAE or PALMAE

2

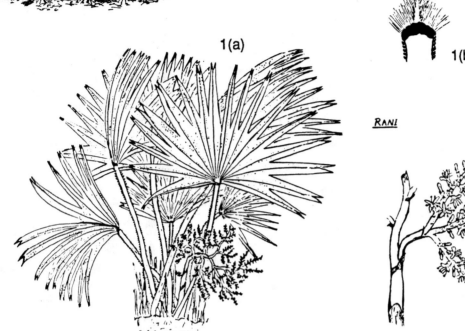

1(a)

1(b)

RANI

IDENTIFYING CHARACTERISTICS (see below for details): 1) leaves — (a) green to blue, deeply divided into stiff segments; (b) 3-sided petioles with recurved teeth on margins of basal half; 2) stem creeps along ground, rarely erect.

DESCRIPTION	Evergreen shrub. Creeps along the ground, rarely erect.
GROWTH RATE	Slow to moderate.
SIZE	Generally 3–8', can reach 25' tall. Trunk 9 to 12" diameter.
LEAVES	Palmate, to 3½' wide, green or blue, deeply divided into 20 or more stiff tapering segments. The 3–4', 3-sided petioles have small, sharp recurved sawteeth covering the margins of the basal half.
FLOWERS	Small, whitish, on 3½'-long, branched flower stalks.
FRUIT	Yellowish, turning black at maturity; ellipsoidal to 1" long; ripens in the summer.
BARK	Covered with old leaf bases and brown fiber.
RANGE	Gulf Coast states, all of Florida and north to the Carolinas.
HABITAT	Wet to dry pinelands and hammocks; dense growth on sandy soils. Tolerant of poor soils if they have adequate drainage.
LANDSCAPE USAGE	Provides naturalistic effect in the landscape. Difficult to transplant so should be left in the landscape during any landclearing.
HUMAN EDIBILITY	Bud used the same as cabbage palm (*Sabal palmetto*), as swamp cabbage, only tastier and using it does not kill the plant as harvesting the heart of cabbage palm does. Ripened fruit was used as a food source by Florida Indians.
REMARKS	Important food source for the threatened Florida Black Bear. Fruit are eaten by other animals and plant is used for nesting and/or cover by wildlife. The flowers are a source of high-grade honey.

POND CYPRESS
(*Taxodium ascendens*)
Family: TAXODIACEAE

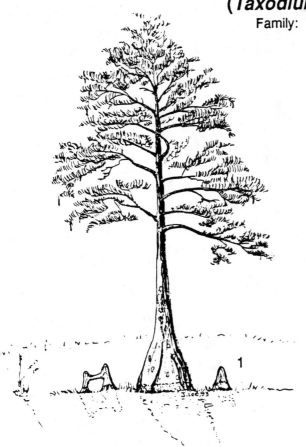

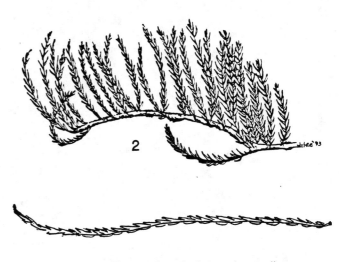

Appressed awl-shaped needles

   Seeds

IDENTIFYING CHARACTERISTICS (see below for details): 1) rounded or blunt-tipped "knees"; 2) deciduous branchlets with ³/₈" scalelike needles.

DESCRIPTION	A medium to large-sized deciduous coniferous tree, with a buttressed base and rounded or blunt-tipped "knees". Branches are horizontal or ascending; top flattens with age.
GROWTH RATE	Moderate.
SIZE	Up to 75–100' and 15–20' wide.
LEAVES	Alternate, simple, smooth, thin scalelike green needles appressed to or spirally arranged on the deciduous branchlets; only up to ³/₈" long, linear awl-shaped, tips pointed. Needles turn coppery-brown in fall.
FLOWERS	No true flowers. Numerous male cones in catkinlike clusters on same tree (monoecious) with female cones which are scattered near the ends of branches.
FRUIT	Seeds borne in cones which are purplish and resinous when young, later becoming greenish-yellow; 1" in diameter.
BARK	Gray to cinnamon-brown, somewhat thin, showing smooth, shallow furrows which become deeper and rougher with age.
RANGE	Southeastern Virginia to southeastern Louisiana, south to Palm Beach County, Florida.
HABITAT	Wet sandy depressions, shallow flatwoods ponds, and pond or lake margins.
LANDSCAPE USAGE	Use as a specimen for large areas such as parks. May be established on drier sites.
HUMAN EDIBILITY	None.
REMARKS	Fruit is eaten by birds. Very resistant to wind damage. Unique conifer since this is the only genus which has deciduous twigs (branchlets). The wood is equally as valuable as that of the bald cypress.

BALD CYPRESS
(*Taxodium distichum*)
Family: TAXODIACEAE

IDENTIFYING CHARACTERISTICS (see below for details): 1) sharp-pointed, conical "knees" around buttressed base; 2) deciduous branchlets with $\frac{1}{2}-\frac{3}{4}$" flat needles arranged in a featherlike pattern.

DESCRIPTION	A large, deciduous, naturally aquatic tree with a buttressed base and sharp-pointed, conical "knees".
GROWTH RATE	Moderate to rapid.
SIZE	Forty to 125' in height; trunks 6' or more in diameter, rarely to 10'.
LEAVES	Alternate, simple, smooth, thin, flat, spreading needles with parallel sides arranged in a featherlike pattern on deciduous branchlets; $\frac{1}{4}-\frac{3}{4}$" long, linear or very narrowly linear-lanceolate. Needles are bright yellow-green in spring, turning to sage-green in summer and russet-brown in fall.
FLOWERS	No true flowers. Numerous male cones in 4–5" drooping clusters on same tree (monoecious) with female cones which are scattered near the ends of branches.
FRUIT	Seeds borne in green to purplish and resinous ball-like cones, brown at maturity (in 1 year); 1" in diameter.
BARK	Gray to reddish-brown, shallowly furrowed and ridged on older trunks, peeling in flaky thin strips.
RANGE	All of Florida, north to Delaware, west to Texas and north to southeast Oklahoma and Illinois.
HABITAT	Wet, swampy soils of riverbanks and floodplain lakes, but is adaptable to well-drained upland soils.
LANDSCAPE USAGE	Specimen or street tree. Can be planted where they add an accent of texture and form.
HUMAN EDIBILITY	None.
REMARKS	Fruit is eaten by birds. Not salt tolerant. Very resistant to wind damage. Commercially harvested for lumber (exterior trim of buildings, planking, boats, docks, posts, poles, etc.). The heartwood is very resistant to decay. Unique conifer, this genus is the only one which has deciduous twigs (branchlets).

FLORIDA ELM
(*Ulmus americana* var. *floridana*)
Family: ULMACEAE

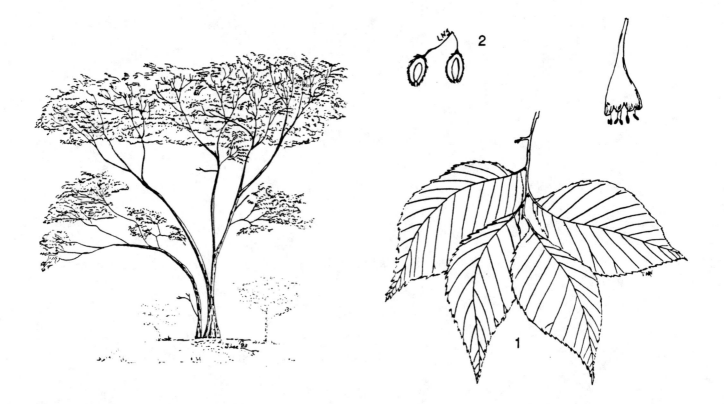

IDENTIFYING CHARACTERISTICS (see below for details): 1) leaves — prominent parallel veins, double-toothed margins; 2) fruit — samaras with deeply-notched tip and hairy edges; 3) twigs — long and smooth, not winged.

DESCRIPTION	Medium-sized deciduous tree with an open, not vaselike, crown.
GROWTH RATE	Moderate.
SIZE	Height of 40–50'; trunk to 18" in diameter. Twigs long and glabrous.
LEAVES	Alternate, deciduous, 2–4" long, 1–2" wide, simple, smooth and shining above, paler and smooth below, elliptic, prominent parallel veins, double-toothed, bases wedge shaped, often unequal. (Less asymmetrical than *U. americana* var. *americana*.)
FLOWERS	Inconspicuous greenish, $\frac{1}{8}$"-wide flowers lacking petals bloom in late winter on slender drooping stalks.
FRUIT	Fruit matures in spring, light green, ovate samara with hairy edges and two-pronged tip; flat seed in center, $\frac{1}{3}$ to $\frac{1}{2}$" long.
BARK	Pale brown to gray, divided by irregular shallow fissures into interlacing flat-topped flaking ridges. (Twigs of *U. alata* and *U. crassifolia* are usually corky-winged.)
RANGE	North Carolina south to DeSoto County in Florida.
HABITAT	Wet to moist hammocks or woodlands and floodplain forests.
LANDSCAPE USAGE	Used as a landscape ornamental; it will grow as a medium-sized ornamental when planted under drier conditions.
HUMAN EDIBILITY	None.
REMARKS	Tough, hard wood has some use as crates and other wood products. We found no reports of Dutch Elm disease reaching central Florida, nearly the southern limit of this tree's range.

SPARKLEBERRY, TREE SPARKLEBERRY
(*Vaccinium arboreum*)
Family: ERICACEAE

IDENTIFYING CHARACTERISTICS (see below for details): 1) small, crooked tree (arboreum) with many branches; 2) leaves — veins prominent on both surfaces; 3) flowers — bell-shaped, blooms later than other *Vaccinium* sp.

DESCRIPTION	Erect shrub to small crooked tree, usually many-branched with slender, rigid, crooked twigs.
GROWTH RATE	Slow to moderate.
SIZE	6–12' average height, to 30'; trunk up to 10" diameter.
LEAVES	Glossy, alternate, leathery evergreen or sometimes tardily deciduous; elliptic to obovate, $\frac{1}{2}$–$2\frac{3}{4}$" long, 1" wide; entire or serrulate, smooth and shiny above, sometimes hairy along veins below; veins prominent on both surfaces.
FLOWERS	Bell-shaped, white corollas on $\frac{3}{4}$–$1\frac{1}{4}$"-long racemes in axils of leaves; (similar looking to lilies-of-the-valley); occur in spring (later than other vacciniums), but over extended period.
FRUIT	Tart fruits mature in mid- to late summer, persist to late winter; shiny, black berry, $\frac{1}{4}$" in diameter.
BARK	Light reddish- to dark brown; covered with thin flaky plates. Twigs — pubescent.
RANGE	North and central Florida, west to southeast Texas, north to Virginia and southeast Kansas.
HABITAT	Sandy uplands, in clearings and open forest understories. Can grow in calcareous soils.
LANDSCAPE USAGE	Background or mass plantings in full sun to partial shade. Tolerant of a wide range of moisture conditions.
HUMAN EDIBILITY	Berries are edible, but not tasty.
REMARKS	Sometimes called tree-huckleberries. Berries are eaten by wildlife. Fire resistant. Close-grained, hard wood is used for tool handles.

HIGHBUSH BLUEBERRY
(*Vaccinium corymbosum*)
Family: ERICACEAE

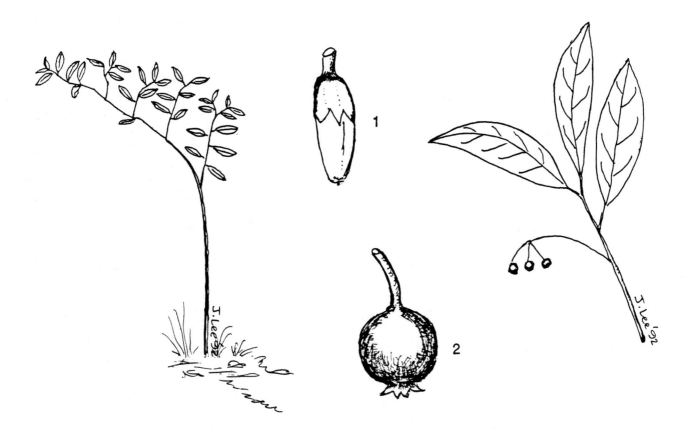

IDENTIFYING CHARACTERISTICS (see below for details): 1) flowers — cylindrical corolla, white or pinkish-tinged; 2) fruit — bluish, glaucous berries, generally larger than those of *V. arboreum*; 3) leaves are over $\frac{5}{8}$" long.

DESCRIPTION	Tall erect deciduous shrub with 1 or more trunks or branches from base.
GROWTH RATE	Slow to moderate.
SIZE	4–10' tall.
LEAVES	Deciduous, glabrous, dark green above and often glaucous; ovate to elliptic, $\frac{3}{4}$–3" long; $\frac{1}{2}$–$1\frac{1}{2}$" wide, margins entire to serrate.
FLOWERS	White or pinkish-tinged, cylindrical, rarely urn- or bell-shaped, $\frac{1}{4}$–$\frac{1}{3}$"-long corolla, in terminal or axillary racemes.
FRUIT	Bluish to blue-black glaucous berry, $\frac{1}{5}$–$\frac{1}{2}$" in diameter.
BARK	Green, yellow or reddish, angular to rounded twigs.
RANGE	Nova Scotia to Illinois, south to central Florida, Texas and Oklahoma.
HABITAT	Moist acidic soils. Pine flatwoods, margins of swamps, and mesic hardwood forests.
LANDSCAPE USAGE	Shrub border or low screen. Yellow, scarlet and crimson fall colors.
HUMAN EDIBILITY	Fruit is excellent out-of-hand or in pancakes and muffins or made into jelly and jam; tea can be made from leaves and flowers.
REMARKS	Wildlife food source.

SHINY BLUEBERRY
(*Vaccinium myrsinites*)
Family: ERICACEAE

IDENTIFYING CHARACTERISTICS (see below for details): 1) size — usually less than 2' tall; 2) flowers — urn-shaped corolla, white to deep pink; 3) stalked glands on lower surface of shiny, less than $5/8$"-long leaves.

DESCRIPTION	A common erect rhizomatous evergreen shrub with a distinct densely-branched growth habit.
GROWTH RATE	Moderate.
SIZE	Up to 2' tall, usually less.
LEAVES	Alternate, glabrous, glossy-green, narrowly oblanceolate to elliptic, up to about $1/2$" long, often with stalked glands on lower, somewhat pale, surface.
FLOWERS	Urn-shaped corolla, white to deep pink; umbel-like clusters or short racemes of 2–8 flowers; present in spring.
FRUIT	Black or glaucous blue berry, $1/4$–$2/5$" in diameter; present in early summer.
BARK	Not applicable.
RANGE	Southeastern U.S., throughout Florida — common in north and central parts of the state.
HABITAT	Acidic pine flatwoods and sandhills.
LANDSCAPE USAGE	Attractive border plant or groundcover.
HUMAN EDIBILITY	Fruit is edible out-of-hand, baked in pancakes and muffins, or made into jelly or jam. Flowers are also edible and tea can be made from dried leaves.
REMARKS	Good wildlife forage.

DEERBERRY
(*Vaccinium stamineum*)
Family: ERICACEAE

IDENTIFYING CHARACTERISTICS (see below for details): 1) flowers — open bell-shaped white corolla with stamens extending beyond bell (stamineum); 2) fruit — variably-colored, frosty-appearing berries; 3) whitish color on underside of leaves.

DESCRIPTION	Small to medium-sized shrub with round twigs so fully branched that they form flat fans.
GROWTH RATE	Moderate.
SIZE	2–12' tall.
LEAVES	Alternate, deciduous, glabrous to pubescent, pale below; 1½–3½" long, elliptic to oblanceolate, entire margins, but gland-tipped teeth near leaf base.
FLOWERS	Corolla white, open bell-shaped, stamens distinctly projecting beyond bell; occurring in racemes. Leafy bracts, smaller than true leaves.
FRUIT	Glabrous to pubescent berry, usually glaucous and tart; variable color, ¼–½" diameter, drops quickly.
BARK	Smooth.
RANGE	Southeastern U.S., common in north and central Florida.
HABITAT	Dry sandy woodlands and hammocks, occasionally in pine flatwoods. Acidic soils.
LANDSCAPE USAGE	Open hedge.
HUMAN EDIBILITY	Fruit edible out-of-hand or made into jelly or jam.
REMARKS	Wildlife food source.

ADAM'S NEEDLE, BEAR GRASS
(*Yucca filamentosa*)
Family: AGAVACEAE

Seed capsule

IDENTIFYING CHARACTERISTICS (see below for details): 1) leaves — basal rosette, margins fray into threadlike fibers; 2) flowers — bell-shaped, creamy white on long stalk.

DESCRIPTION	Herbaceous clump-forming perennial.
GROWTH RATE	Slow.
SIZE	Up to about 2½'; flower stalk 3–10' tall.
LEAVES	Basal rosette, swordlike but not stiff, 1–2½' long, with margins fraying into coarse threads.
FLOWERS	Creamy white, bell-shaped, 1–3" in diameter, clustered in panicles at end of long stalk (up to 16'); bloom in spring, branches of inflorescence glabrous.
FRUIT	Capsules, on erect inflorescence, 1½–2" long, ¾" in diameter.
BARK	Not applicable.
RANGE	South in Florida to Palm Beach County, north to Virginia, west to Mississippi.
HABITAT	Sandhills and dry open woodlands.
LANDSCAPE USAGE	Water-conserving landscapes, planters, naturalistic landscapes.
HUMAN EDIBILITY	Only petals are eaten out-of-hand or in salads, other flower parts are bitter.
REMARKS	Leaf fibers were used by Indians as cordage (rope, string, etc.); roots used as a shampoo and soap. Pollinated only by yucca moth (*Tegeticula alba*).

COONTIE, FLORIDA ARROWROOT
(*Zamia floridana, Z. pumila*)
Family: CYCADACEAE or ZAMIACEAE

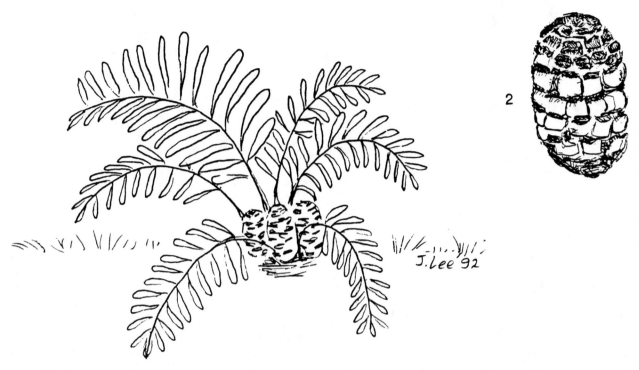

2

J. Lee '92

IDENTIFYING CHARACTERISTICS (see below for details): 1) leaves — 2–4' long, pinnately compound, stiff; 2) fruit — bright orangish-red fleshy, borne on cones at ground level.

DESCRIPTION	Short, fernlike evergreen shrub.
GROWTH RATE	Slow.
SIZE	3' tall.
LEAVES	Terminal crown of underground stem; stiff, evergreen, palmlike, to about 3' long; pinnae up to 6" long and $\frac{1}{2}$" wide.
FLOWERS	Not applicable. The gymnosperms bear naked seeds, usually on the scale of a cone (see next item).
FRUIT	Large, reddish-brown fleshy cones at ground level. Dioecious. Male pollen-producing cones are smaller ($\frac{1}{2}$–$1\frac{1}{2}$" by 4" long) than female cones (2" by 6" long). Seeds on mature female cones are orange to red and angular.
BARK	Not applicable. (Stem is short or underground, tuberlike.)
RANGE	Throughout Florida.
HABITAT	Tolerant of various well-drained soils of hammocks and pinelands.
LANDSCAPE USAGE	Used in the landscape for foundation plantings and in planters; as a specimen or transition plant and as a groundcover. Tolerant of full sun but does better in shade. Tolerant of salt spray.
HUMAN EDIBILITY	Root very high in starch, but hydrocyanide **must** be leached out.
REMARKS	Listed as rare, threatened or endangered; protected by law. Seminole Indians called this plant *Contihateka,* which means white-root or white bread-plant, and used the rootstock as a food source (see edibility).

Word Glossary

acid (soil) - Technically a soil with a pH (hydrogen ion concentration) value <7.0; however, plants listed as requiring acid soils for good growth (for example *Vaccinium* sp.) usually do best in soils with pHs between 4.5 and 5.5.

alternate - Attached singly at different levels along an axis as with leaves or buds attached one per node to a stem. Arrangement of parts not opposite or whorled.

apex - Tip or terminal end.

appressed - Pressed or lying flat or close to something, for example, scalelike leaves of *Juniperus* (southern red cedar).

awl-shaped - Tapering from a broader base to a slender sharp point.

axil - Upper angle between a stem and another plant part, such as a leaf or branch.

berry - Stoneless, fleshy fruit consisting of many individual seeds.

boot - Trunk-end of a palm frond which may remain attached after the rest of the frond has detached and fallen.

bract - A leaf, often reduced in size, which is just below a flower or inflorescence; often remaining under the fruit also. In conifers, the modified leaf or scale attached to the central axis of the cone.

calcareous (soil) - Soils containing significant amounts of lime (calcium carbonate).

calyx - Collective term for all the sepals of a flower; especially used when sepals are of a different size (usually smaller) or color (often green) than the petals.

canopy - The uppermost layer of foliage in a plant community.

capsule - A fruit composed of two or more parts that dries and opens and releases seeds.

compound (leaf) - A leaf made up of two or more leaflets.

cone - A structure composed of modified leaves (scales or bracts) which bear the reproductive organs of gymnosperms; the female structure ultimately leads to the development of seeds that are not enclosed, thus, they are called "naked seeds". (A cone is also called a strobilus.)

conifer - A cone-bearing tree or shrub.

coniferous - Cone-bearing.

corolla - Collective term for all the petals of a flower.

crown - The uppermost portion of the foliage of a plant. Also used to refer to the persistent base of a perennial herb from which new growth arises. (In addition, refers to some appendages of flower parts.)

cultivar - A cultivated variety (variation) of a plant; generally selected for specific characteristics.

cyme - A flat-topped, several-branched inflorescence with the central terminal flowers opening before the outer flowers; for example, *Sambucus* (elderberry). (See Picture Glossary.)

deciduous - Plants that drop all their leaves at the same time leaving the tree leafless for a period of time; leaves often fall in autumn each year. (For other plant organs, dropping before most of the related organs.)

dioecious - Bearing male and female reproductive organs (flowers or cones) on separate plants.

drupe - A fruit with a fleshy or pulpy mass surrounding an inner hard cover generally enclosing a single seed; for example, *Prunus* (peach).

evergreen - Plants that do not lose all their leaves at one time.

fascicle sheaths - Tubular structures around the base of a cluster of leaves (needles); for example, *Pinus* sp. (pines).

flatwoods - Level low-lying habitats with acid soil underlain by clay or hardpan; the most common plant community in Florida.

flower - The blossom or inflorescence; generally the part of a plant that bears reproductive organs — male, female or both — and which, if female, ultimately leads to the development of enclosed seeds (as opposed to the naked seeds of gymnosperms).

glabrous - Without hairs.

gland - A structure on the surface of an organ (e.g. leaf) which secretes or stores a fluid.

glaucous - Covered with a thin, whitish, often waxy, coating which can be readily rubbed off.

globose - Rounded or spherical in shape.

gymnosperm - A plant that bears seeds that are naked (not enclosed in a fruit).

hammock - An area, usually higher in elevation than its surroundings, characterized by hardwood vegetation on rich, fertile soil.

hardwood - Broad-leaved, flower-bearing trees — as opposed to coniferous ones; may or may not have hard wood.

herbaceous - Plants having little or no woody tissue.

hydric - Related to an abundance of moisture.

inflorescence - The flowering part of a plant; the arrangement of flowers on a plant (see Picture Glossary).

lenticel - A pore or corky area on young stems or trunks of woody plants.

lichen - An organism composed of alga and fungus living symbiotically (mutually beneficial). Variously colored and flat, flaky, crusty or spongy in appearance.

lobe - A part of a leaf or other organ that is partly divided, forming a projecting part of the organ.

margin (leaf) - The edge of the leaf.

mesic - Moderately moist.

midrib - The central vein of a leaf.

monoecious - Bearing male and female reproductive organs (flowers or cones) on the same plant.

naval stores - Resinous products, such as turpentine and rosin, obtained from conifers such as pines.

node - A location on a stem or branch where new structures (e.g. leaf, stem or flower) originate.

nutlet - A small one-seeded nut or a fruit similar to a nut.

opposite - Attached in pairs at the same level (node) on a stem, as with leaves or buds. Arrangement of parts not alternate or whorled.

overstory - The uppermost layer of foliage in a plant community, formed by the tallest plants. This layer modifies the microclimate (light, humidity, etc.) and thereby helps determine the number of other layers and kinds of plants which live below it.

ovoid - Egg-shaped.

pedicel - The stalk of an individual flower in a flower cluster.

peduncle - The stalk of a flower cluster, solitary flower or a fruit.

perennial - A plant living for three or more years.

petal - A leaflike structure of a flower; usually showy and colorful. Collectively, the petals of a flower are called the corolla.

petiole - The stalk of a simple leaf or the primary stalk of a compound leaf.

pH - The number expressing the acidity or alkalinity of a substance. Numbers above 7 indicate alkalinity, while numbers below 7 indicate acidity.

pinna (*pl.* pinnae) - A leaflet or primary division of a pinnately compound leaf. (See Picture Glossary.)

pistil - The female reproductive parts of a flower.

pistillate - Female flowers; bearing pistils, but no functional stamens.

prairie - Treeless, or nearly treeless, flatlands covered by grasses and other low plants.

pubescent - With hairs.

rachis - The stem of an inflorescence or the leaflet-bearing extension of the petiole of a compound leaf.

revolute - Leaf margins curved downward.

rhizomatous - Having a rhizome.

rhizome - A horizontal stem, usually underground and spreading.

rosette - A circular arrangement of leaves at the base of a plant.

samara - A winged fruit, usually one-seeded; for example, *Acer* (maple).

scale - A modified leaf (bract) that bears the naked seeds in a cone. Also, thin, flaky particles, often nearly translucent, occurring on stems or the undersides of leaves.

scrub - Areas characterized by deep, excessively well-drained sandy soils. Due to the low water-holding capacity of the soils, conditions are similar to those of a desert much of the time; therefore, the plants inhabiting these areas are adapted to xeric conditions.

sepal - One of the outermost set of flower parts, just outside the petals; usually green and resembling a small leaf. Collectively, the sepals of a flower are called the calyx.

sessile - Without a stalk; for example, a leaf (with no petiole) attached directly to a branch.

simple (leaf) - A leaf, not composed of leaflets, but which may be deeply lobed.

stalk - Slender supporting structure of an inflorescence or fruit (peduncle), or a leaf (petiole).

stamen - Male reproductive parts of a flower.

staminate - Male flowers; bearing no functional pistils.

striated - Longitudinally streaked, ridged, grooved, furrowed or striped.

subcanopy - The plants comprising the understory of a plant community; adapted to the lower light levels under the canopy.

subsessile - Short-stalked, almost stalkless.

swamp - Areas inundated with water at least part of the year, cypress and bay are often the dominant tree species in freshwater swamps in Florida.

terminate - At the end of a structure.

tuber - A fleshy enlarged portion of a stem or root, generally used for storage, which may be used for vegetative reproduction.

understory - Layer composed of plants tolerant of reduced light and growing beneath the overstory plants.

veins - Strands of (vascular) tissue used to transport substances through plants.

venation - The arrangement of the veins (vascular/transport system) of a leaf.
    palmate - Radiating from a common point, like the fingers on a hand.
    parallel - Extending in the same direction and about equidistant apart along the leaf.
    pinnate - Lateral veins arising from a central axis; featherlike pattern.

whorled - Three or more leaves, branches or flower parts arranged in a circular pattern around a common axis.

xeric - Very dry or relating to a lack of, or low, moisture.

Bibliography

Angier, Bradford. 1974. **Field Guide to Edible Wild Plants**. Stackpole Books, Harrisburg, PA.

Bailey, Liberty H. and Ethel Z. Bailey. 1976. **Hortus Third: A Concise Dictionary of Plants Cultivated in the United States and Canada**. MacMillan Publishing Co., New York, NY.

Baker, Mary F. 1938. **Florida Wild Flowers, An Introduction to the Florida Flora**. The MacMillan Company, New York, NY.

Bell, C. Ritchie and Bryan J. Taylor. 1982. **Florida Wild Flowers and Roadside Plants**. Laurel Hill Press, Chapel Hill, NC.

Brockman, C. Frank. 1968. **Trees of North America**. Golden Press, Western Publishing Company, New York, NY.

Cerulean, Susan, Celeste Botha, Donna Legare and Swannee Nardandrea. 1986. **Planting a Refuge for Wildlife, How to create a backyard habitat for Florida's birds and beasts**. Florida Game and Fresh Water Fish Commission Nongame Wildlife Program and U. S. Department of Agriculture Soil Conservation Service.

Chanticleer Press, Inc. 1988. **Taylor's Guide to Trees**. Houghton Mifflin Co., Boston, MA.

Clewell, Andre. 1985. **Guide to the Vascular Plants of the Florida Panhandle**. University Presses of Florida, Tallahassee, FL.

Deuerling, Richard J. and Peggy S. Lantz. 1993. **Florida's Incredible Wild Edibles**. Florida Native Plant Society, Orlando, FL.

Dirr, Michael A. 1983. **Manual of Woody Landscape Plants: Their Identification, Ornamental Characteristics, Culture, Propagation and Uses**. Stipes Publishing Co., Champaign, IL

Dressler, Robert L., David W. Hall, Kent D. Perkins and Norris H. Williams. 1987. **Identification Manual for Wetland Plant Species of Florida**. University of Florida, Institute of Food and Agricultural Sciences, Florida Cooperative Extension Service, SP-35.

Elias, Thomas S. and Peter A. Dykeman. 1982. **Field Guide to North American Edible Wild Plants**. Outdoor Life Books, New York, NY.

Fleming, Glenn, Pierre Genelle and Robert W. Lang. 1984. **Wild Flowers of Florida**. Banyan Books, Inc., Miami, FL.

Florida Committee on Rare and Endangered Plants and Animals. 1979. **Rare and Endangered Biota of Florida. Vol. 5. Plants**. Daniel B. Ward, Editor. University Presses of Florida, Gainesville, FL.

Florida Department of Agriculture, Division of Forestry. 1985. **Forest Trees of Florida**. Tallahassee, FL.

Florida Department of Agriculture, Division of Forestry. 1980. **Urban Trees for Florida**. Tallahassee, FL.

Foote, Leonard E. and Samuel B. Jones, Jr. 1989. **Native Shrubs and Woody Vines of the Southeast**. Timber Press, Inc., Portland, OR.

G. & C. Merriam. 1956. **Webster's New International Dictionary of the English Language**. Unabridged. 2nd edition. William Allan Neilson, Editor in Chief. G. & C. Merriam Company, Springfield, MA.

Gibbons, Euell. 1968. **Stalking the Wild Asparagus**. David McKay Company, Inc., New York, NY.

Godfrey, Robert and Jean Wooten. 1981. **Aquatic and Wetland Plants of Southeastern United States**. The University of Georgia Press, Athens, GA.

Harlow, William and Ellwood Harrar. 1969. **Textbook of Dendrology**. McGraw-Hill, New York, NY.

Harrington, H. D. 1979. **How to Identify Plants**. Ohio University Press, Athens, Ohio.

Hume, H. Harold. 1953. **Hollies**. The MacMillan Co., New York, NY.

Kurz, Herman and Robert K. Godfrey. 1962. **Trees of Northern Florida**. University of Florida Press, Gainesville, FL.

Little, Elbert L. 1978. **Atlas of United States Trees**. Misc. Pub. No. 1361, U.S. Government Printing Office, Washingtion, D.C.

Little, Elbert L. 1987. **The Audubon Society Field Guide to North American Trees — Eastern Region**. Alfred A. Knopf, Inc., New York, NY.

Morton, Julia F. 1977. **Plants Poisonous to People in Florida and Other Warm Areas**. Fairchild Tropical Garden, Miami, FL.

Ornamental Horticulture Department, Institute of Food and Agricultural Sciences, University of Florida. 1984. **Ornamental Horticulture Plant Identification Manual, Volume 1**. Gainesville, FL.

Ornamental Horticulture Department, Institute of Food and Agricultural Sciences, University of Florida. 1984. **Ornamental Horticulture Plant Identification Manual, Volume 2**. Gainesville, FL.

Petrides, George A. 1972. **A Field Guide to Trees and Shrubs**. Houghton Mifflin Company, Boston, MA.

Radford, Albert E., Harry E. Ahles and C. Ritchie Bell. 1979. **Manual of the Vascular Flora of the Carolinas**. The University of North Carolina Press, Chapel Hill, NC.

Rogers, David J. and Constance Rogers. 1991. **Woody Ornamentals for Deep South Gardens**. University of West Florida Press, Pensacola, FL.

Small, John Kunkel. 1933. **Manual of the Southeastern Flora**. The University of North Carolina Press, Chapel Hill, NC.

Tarver, David P., John A. Rodgers, Michael J. Mahler and Robert L. Lazor. 1979. **Aquatic and Wetland Plants of Florida**. Florida Department of Natural Resources, Tallahassee, FL.

Taylor, Walter Kinglsey. 1992. **The Guide to Florida Wildflowers**. Taylor Publishing Company, Dallas, TX.

Van Atta, Marian. 1985. **Wild Edibles Identification for Living Off the Land**. Pine & Palm Press, Melbourne, FL.

Watkins, John V. and Thomas J. Sheehan. 1975. **Florida Landscape Plants, Native and Exotic**. The University Presses of Florida, Gainesville, FL.

West, Erdman and Lillian E. Arnold. 1956. **The Native Trees of Florida**. University of Florida Press, Gainesville, FL.

Whitcomb, Carl E. 1985. **Know It and Grow It, II: A Guide to the Identification and Use of Landscape Plants**. Lacebark Publications, Stillwater, OK.

Wunderlin, Richard P. 1982. **Guide to the Vascular Plants of Central Florida**. University Presses of Florida, Gainesville, FL.

Index

Items listed in light type are not described in detail, but are mentioned on the page listed.

Common Native Plants

# of Central Florida

(Lake, Orange, Osceola
and Seminole Counties)

Tarflower Chapter
Florida Native Plant Society
P.O. Box 564
Orlando, Florida 32802